FIRE

OCCUPATIONAL SAFETY AND HEALTH

A Series of Reference Books and Textbooks
on Occupational Hazards • Safety • Health •
Fire Protection • Security • and Industrial Hygiene

Series Editor
ALAN L. KLING
Loss Prevention Consultant
Jamesburg, New Jersey

Other Volumes in Preparation

FIRE
Fundamentals and Control

Walter M. Haessler
Santa Rosa, California

CRC Press
Taylor & Francis Group
Boca Raton London New York

CRC Press is an imprint of the
Taylor & Francis Group, an **informa** business

CRC Press
Taylor & Francis Group
6000 Broken Sound Parkway NW, Suite 300
Boca Raton, FL 33487-2742

First issued in paperback 2020

© 1989 by Taylor & Francis Group, LLC
CRC Press is an imprint of Taylor & Francis Group, an Informa business

No claim to original U.S. Government works

ISBN 13: 978-0-367-58024-7 (pbk)
ISBN 13: 978-0-8247-8024-1 (hbk)

**Visit the Taylor & Francis Web site at
http://www.taylorandfrancis.com**

**and the CRC Press Web site at
http://www.crcpress.com**

Preface

The general area of fire protection has been well covered by many publications. *The Fire Protection Handbook* of the National Fire Protection Association (NFPA) is the most comprehensive in size and scope; even in that book there are limits to the amount of detail and coverage.

It is my intent to present the admittedly complex phenomenon of combustion in as simple and straightforward a manner as possible. In the interest of simplicity, redundant material that can be easily obtained from present literature will not be included. The ample bibliography will satisfy the needs of most readers.

In a general way, the Table of Contents for this text lists sequentially the various factors in the combustion process. These factors are illustrated diagrammatically in the figure on the following page. It will be seen that there are two modes by which fire reveals itself; the glowing and the flaming modes, which are notably different from each other. There is an interplay of endothermic reactions (energy-absorbing) and exothermic reactions (energy-emitting); the latter are predominant, with the net result that the combustion process emits heat.

These factors are highly temperature-dependent and occur only under temperature conditions in excess of 1500°F (815°C) and are approximately visualized as a bright red heat. Interestingly,

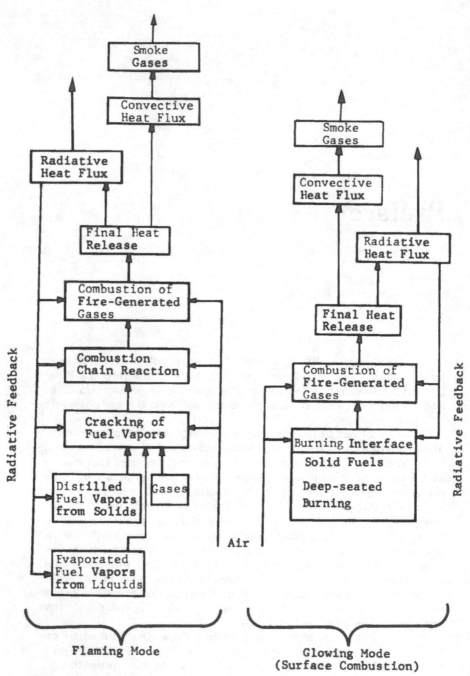

The two modes of combustion.

the NFPA has defined a combustible material as that which can be exposed, in air, to 1500°F for 5 min with no evidence of burning. (The word *combustion* is derived from Latin and is defined as the past participle of the verb *combustus* which, translated, means *burnt up*.)

As important as it is to view these factors separately, we must realize that they are not mutually exclusive but, rather, they over-lap, sometimes to a considerable extent. The terms *flammable* and *combustible* have, at times, become confused. Flammability ratings have been devised that classify the relative ease of ignition of gases and vapors derived from volatile liquids versus the heavier, low-volatile liquids and the solids that present a greater resist-ance to ignition.

This text is intended for use by students, engineers, safety directors and inspectors, fire insurance personnel, and members of various governmental, public, and private fire services.

The reader must, to some extent, be familiar with the general subject and have a basic understanding of the physical sciences and how they relate to the subject.

It has always been my contention that theory and practice should complement each other to achieve a full understanding of the many factors involved and the influences that they exercise. Thus, future trends can be anticipated for promoting the efficiency of fire protection and improving the techniques of fire combat. The importance of improved fire protection engineering cannot be over-emphasized, for it intersects all fields of human endeavor. The fire prevention engineer must truly be a "man for all seasons."

Walter M. Haessler

Introduction

Ever since man's monumental discovery of fire as a means of making his existence more tenable, major attention has been directed toward its beneficial use. The key to this benefit is control. A great amount of information has been accumulated toward this end; in fact, most of what we know about fire concerns the controlled burning of fuels. It was portentous that the principles of modern chemistry were discovered largely as a result of experiments conducted using controlled combustion, by the eminent French scientist Lavoisier about 200 years ago. When this control is lost, the fire becomes wild and considerable effort usually will be required before control is regained. Our knowledge of fire extinguishment is comparatively recent, so where do we stand?

From ancient times to the advent of the Industrial Revolution, only about 150 years past, our culture rested upon stone and wood. Urban conflagrations became more frequent as the population increased, and fire protection was based solely on the use of water and the availability of manpower.

As the Industrial Revolution progressed, steam-driven water pumps increased the efficiency of water. Hand-portable and soda/acid-type water extinguishers became available about the time of the American Civil War.

With the advent of petroleum and its derivatives, not only did fire hazards increase but the nature of the fire was changed. Chemical foam was developed and was dispensed with a modified version of the soda/acid-type water extinguisher. The increasing use of automobiles, small boats, and aircraft resulted in the development of portable carbon tetrachloride pump-type extinguishers. Dry chemical extinguishers showed promise, but vexing mechanical difficulties existed.

The foregoing was the general state of the art by the end of World War I, when carbon dioxide was employed in the first major developments in fire protection: hand-portable extinguishers and installed stationary systems. Marine fire protection became a reality, and safety at sea was greatly enhanced. By a clever use of common piping, an automatic smoke-detection system could be included with the carbon dioxide piping.

Mechanical protein-based and synthetic-based foams were developed to combat large-scale oil fires on land and at sea in the period between World Wars I and II. An American admiral was credited with the comment that the war in the Pacific could not have been won without the use of fire-fighting foam.

The mechanical difficulties that plagued the dry chemical extinguishers were gradually overcome, and a new dimension to fire fighting was accomplished. Various dry chemical formulations, together with apparatus ranging from small portable extinguishers to large mobile and stationary systems, greatly extended fire-fighting capabilities.

Just before World War II, two new halogenated hydrocarbons were developed that were markedly superior to carbon tetrachloride and provided a means of achieving aircraft fire protection within reasonable weight limitations. The British developed methyl bromide systems for their Spitfires, and the Germans developed bromochloromethane for their Messerschmitts. These agents were highly toxic. In our country, research developed three new halogenated hydrocarbons that were improved from the standpoints of both effectiveness and reduced toxicity.

After World War II, the technology of fire control was greatly expanded and progress was so prolific that it now becomes necessary to itemize it.

1. Automatic water sprinkler systems, which had been used for at least 100 years, were greatly improved and diversified. When properly installed and maintained, their superlative performance is now unquestioned.

2. Newer and more effective dry chemical formulations and equipment were devised for all types of fires (classes A, B, C, and D) ranging from small wood fires to large fires involving combustible metals.

3. Newer, less toxic, and more effective halogenated hydrocarbons were developed to protect against hazards from which previously no protection could be accomplished

4. Additives for water were developed that would either emulsify flammable liquid fuels or prevent embering.

5. Special slurries and solutions were developed for use from aircraft in support of ground crews combatting wildfires.

6. One stunning development was the ability to anticipate and extinguish an explosion, after ignition, and prevent the devastation of a detonation.

As will be seen, the tempo of improvements in materials and techniques has rapidly increased, and there is no reason to believe that it will slacken. The arsenal available to both fire fighters and fire protection engineers is constantly increasing, requiring greater knowledge and sophistication in the selection of an ever-increasing number of choices. Specialization will result in more efficient use of human and material resources, integrated with economic limitations. There is no doubt that the future of fire control will undergo considerable adaptation.

It could well be that the future will mean the development not only of more separate attacks on fire but, in addition, of a more intelligent approach utilizing the various available extinguishing agents and techniques in concert and in an optimal manner. This calls for an understanding of the basic scientific knowledge of fire and its behavior and how it is influenced by the type of fuel, the degree of ventilation, the type of oxidant, and the effect of the generated heat energy on the fire itself as well as on the surrounding environment.

For example, this has become particularly apparent in suburban and rural areas where the water supply is not inexhaustible. These areas are larger than their urban counterparts and are more limited in human and material resources. When the tax-paying public starts to cut the budgets of fire departments, the results are fewer personnel, with less equipment to cover relatively large areas with ever-increasing populations. Furthermore, the water supply is usually additionally taxed.

The foregoing example, together with many more, calls for more education and training in the use of newer techniques that

employ not necessarily only one extinguishing agent but instead
use a synergistic multiagent approach. Mobile fire apparatuses
will no doubt be due for considerable changes.

It is unquestioned that the technology of fire control is complex: it intersects all the basic disciplines of the four branches of
engineering: civil, electrical, chemical, and mechanical.

Contents

FIRE

1
Modes and Types of Combustion

This chapter is an introductory opening to a particularly broad subject and cannot be considered separately from the chapters that follow. The subject matters of these chapters overlap considerably so that the various entities can be knitted into an intelligible whole. The coverage must, perforce, be of a general nature and yet be sufficiently detailed to stimulate the reader to further study. The inclusion of numerous illustrations will be of additional assistance.

The combustion process is observed as occurring in either of two modes: the flaming mode or the glowing mode.

THE FLAMING MODE

The flaming mode is a characteristic indicative of the free burning of either gases or vapors that are derived indirectly from liquid or solid fuels or directly from flammable gases. The latter require no preparation to enter the burning process. Flammable liquids, as such, do not enter the burning process directly because evaporation is required to form the flammable vapors. The heat energy to accomplish this is derived from the heat re-

leased by the fire itself in the form of radiative feedback. With
flaming solid fuels, vapors are derived from the pyrolytic decomposition through the effect of radiative feedback from the
fuel itself. As seen from the frontispiece, radiative feedback
also motivates other phases of the combustion process such as
the "cracking" of complex fuel vapors, thus forming free radicals as well as simpler vapors. Similarly, radiative feedback
stimulates the interplay of chain reactions, as well as the combustion of fire-generated gases. The intensity of any combustion process is always expressed as the time rate of heat energy
released to the surrounding environment. In terms that we are
familiar with, we would express this rate as British thermal
units per minute (BTU/min).

Liquid fuels in the form of mists or sprays, and solid fuels
in the form of dusts or pulverized materials, flame in a manner
similar to liquid fuels. Volatility and large particle surface area/
mass ratios are responsible for the similarity, because all combustion processes occur at the interface between the fuel and
the air.

After ignition, the pace of burning accelerates exponentially.
Combustion is a display of the conversion of chemical energy,
locked up within the fuel molecules, to heat energy. In the
early stage of the fire, the chemical reactions are energy absorbing (endothermic), and these are followed by energy-emitting (exothermic) reactions. The latter reactions emit higher
rates of heat energy release than that absorbed by the endothermic reactions. It is this characteristic that gives the fire
its self-sustaining nature, providing that all of the factors
shown in the frontispiece are not interfered with. As a result
of the foregoing, the increasing rate of heat release will cause
the fire to expand until it reaches the limits of its environment,
it ceases for lack of fuel and air, or it ceases because of extinguishment.

It is noteworthy that the theoretical flame temperatures for
the burning of gases in air, under conditions of no excess air,
do not vary appreciably with different gases, despite the large
differences in heats of combustion. For example, hydrogen,
with a heating value of 319 BTU/ft^3 at 60°F and 1 atm, has a
theoretical flame temperature of about 3960°F, whereas benzol
with a heating value more than 11 times as great, or 3744 BTU/
ft^3, gives a temperature of 3850°F. This is because gases with
the higher heating value (larger molecules) also require correspondingly larger amounts of air for combustion. Most calcu-

lated flame temperatures for complete combustion in air, with no excess air and no heating of the air or gas, lie between 3500 and 4200°F. Higher temperatures can be attained only by pre-heating, or by using oxygen-enriched air, or by both.

However, under hostile fire conditions, the bulk of experimental evidence shows that flame temperatures range from 2300 to 2500°F because the fire is underaerated, causing incomplete combustion and heavy smoke production. The rate of emitted heat flow will increase until a limit is reached. In the fire testing of manual, portable fire extinguishers, square pans or pits are used with a liquid, flammable fuel. The fire areas have been standardized and range in size from 5 to 1200 ft^2. In all of the trials, the fire spreads and grows rapidly from the source of ignition. The rate of heat emitted will, at first, be limited by the size of the firepan or pit, and will shortly, thereafter, increase in intensity until steady-state conditions prevail. At this point, the rate of heat energy released is balanced by the rate at which heat is being lost to the surrounding environment. The characteristic orange yellow color associated with the flames from carbon-bearing fuel gases and vapors is due to the presence of free incandescent carbon in the flaming area, resulting from the "cracking" of complex fuel molecules as well as from a deficiency of air.

THE GLOWING MODE

The glowing mode is exhibited in the form of surface combustion in which the burning is localized to the interfacial surface between the fuel and the air. Pure examples of fuels that burn according to the glowing mode include numerous metallic and nonmetallic chemical elements, from which gases and vapors are not distilled and no flaming exists. Examples of these elements are

Magnesium	Zirconium
Aluminum	Uranium
Lithium	Plutonium
Sodium	Zinc
Potassium	Phosphorus
Titanium	Boron

Their incendiary properties vary over a wide range. White phosphorus emits a ghostly greenish light when viewed in a

dark room and, if the air is humid, autoignition will occur at
86°F. Zirconium, uranium, and plutonium, despite their relatively high ignition temperatures when in solid form, when in
the form of particles can spontaneously ignite in moist air. The
range of properties of these materials is so wide that no complete statement can be made for all, with the exception that the
smaller the particles the more easily it is for ignition to occur.
Temperatures produced by burning metals are generally much
higher than temperatures generated by burning flammable liquids. Some hot metals continue burning in carbon dioxide, nitrogen, or steam atmospheres in which ordinary combustibles or
flammable liquids would be incapable of burning. To apply the
correct type and quantity of an extinguishing agent, and in the
right manner, becomes a highly selective matter.

Glowing and smoldering fires, such as exist with charcoal and
coke, are more familiar. As previously stated for wood, pyrolytic decomposition and the resultant distillation of flammable vapors, including methane, methyl alcohol, formic anhydride, formic acid, and acetic acid, when combined with free water vapors,
give the fire its flaming character. As the fire progresses, char
begins to form, and the transition continues until the original
wood has become charcoal, which is elemental carbon plus residual ash. Here, we now have, as shown in Figure 1.1, an example of wood burning originally in the flaming mode, making
its transition to the final glowing mode. The combustion of coal
follows a similar pattern, with coke as the end product.

Another characteristic of burning charcoal and coke is the
appearance of small bluish, nonluminous flames accompanying the
red glow in the later stages of burning. In the earlier stages
of the fire, the orange yellow color of the flames and their radiative nature are due to the presence of elemental carbon particles that are formed in the combustion reactions. Of all the
various burning species within the flames, carbon burns the
most slowly and participates in a number of steps that, if interfered with, cause increased evolution of carbon monoxide and
smoke. As the fire subsides to its glowing stage, all of the
other volatile fuel elements have been burned away leaving the
slow-burning carbon. The unique existence of the short, bluish, nonluminous flames, which result from carbon monoxide
burning to carbon dioxide, is important in understanding the
twin phenomena that are experienced in enclosed spaces and are
termed *flashover*, and *backdraft*. These will be covered later
in Chapter 5 on fire-generated gases.

There are two forms whereby the flame mode is carried out.

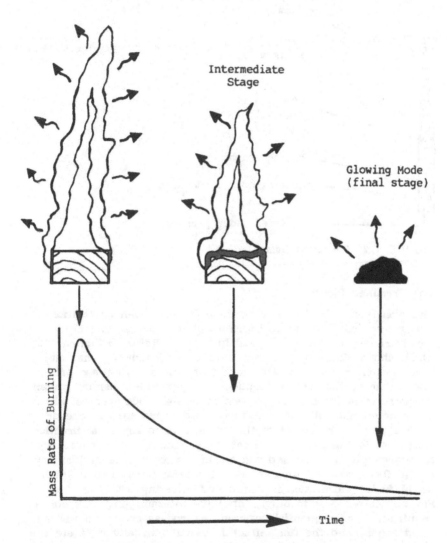

FIGURE 1.1 Flaming and glowing combustion.

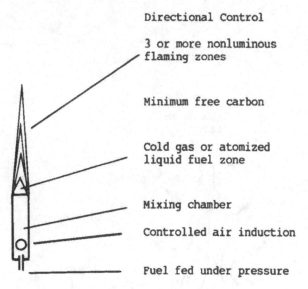

Directional Control

3 or more nonluminous
flaming zones

Minimum free carbon

Cold gas or atomized
liquid fuel zone

Mixing chamber

Controlled air induction

Fuel fed under pressure

FIGURE 1.2   Premixed flame.

The Premixed Flame

The premixed flame is the type used in utilitarian applications
that range from the simple Bunsen burner, to gas and liquid
fuel burners, to pulverized coal burners. Refer to Figure 1.2,
which shows the simplest example of a gas burner. The same
basic principle holds for all types of burners. In these premixed flames, the fuel is supplied in a controlled manner under
pressure from the air supply, which is also under control whether inducted naturally or introduced under pressure in one or
more stages. The object of the process is to impart a directional control to the flames as dictated by design. It is necessary
to intermingle the fuel and the air in a mixing zone preliminary
to the flame zone. With gas, this is easily accomplished; with
liquid fuels, atomization is required for mixing with the air.
For pulverized coat to burn, intensive intermingling with air is
required. The sequential combustion process occurs in well-de-
fined zones, and the final effect is to substantially complete the
burning at the flame tip, at which location the flame temperature is at a maximum. The design accomplishes this by closely
matching the fuel and airflow with as little excess air as possible to achieve smokeless combustion.

The Diffusion Flame

The diffusion flame occurs in open, free-burning, so-called hostile situations such as gas, flammable liquid, and solid fuel fires, and whenever no control over the rate of burning can be exercised other than containment or extinguishment. As contrasted with the premixed flame, there is no external control of either the fuel or the air supply. From Figure 1.3, it is apparent that the fire naturally inducts air by means of its own thermal updraft, which diffuses into the flames from the surrounding environment. When the air enters at a temperature of 100°F and quickly rises to temperatures in excess of 2000°F, it undergoes an expansion exceeding a 4:1 ratio, thereby creating a low-pressure region within the fire. This causes the air to rush in at velocities dependent upon the total time rate of heat release accompanied by the formation of "twisters." On a catastrophic scale, the huge fires that destroyed cities during World War II literally altered the prevailing meteorological conditions, thereby creating a firestorm.

This manner of air supply always results in inadequate aeration that results in incomplete combustion and the formation of poisonous gases and smoke, which are carried off by the convective heat flux (see Preface, p.iv). A substantial portion of the radiative heat flux is directed back toward the fire which provides the stimulus for the fire to expand and, thus, has been defined as *radiative feedback*. This term will be used frequently in this text. Hence, the net exothermal nature of the fire makes it feed itself with evaporated or distilled fuel vapors and air, uncontrolled, and ever-expanding until some environmental limit is reached, or until the fuel is consumed or the fire is extinguished.

In referring to the frontispiece, it will be noticed that for burning fuel gases, *no* radiative feedback is required because the fuel does not need to be evaporated or distilled in preparation for burning. The only controlling factors are

1. The size of the opening of ruptured pipes, tanks, valves, flanges and associated equipment.
2. The prevailing pressure under which the gas is flowing to the atmosphere.
3. The specific gravity of the gas.
4. The volumetric heat content of the gas.
5. If liquefied gas is escaping, the relatively higher density of the liquid phase, compared with the gaseous

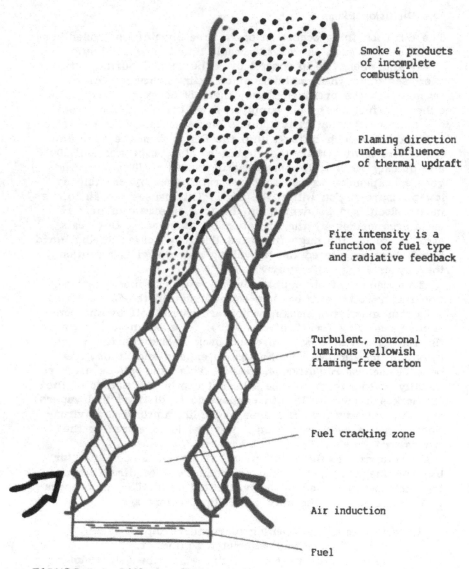

Smoke & products
of incomplete
combustion

Flaming direction
under influence
of thermal updraft

Fire intensity is a
function of fuel type
and radiative feedback

Turbulent, nonzonal
luminous yellowish
flaming-free carbon

Fuel cracking zone

Air induction

Fuel

FIGURE 1.3 Diffusion flame.

phase, becomes a matter of critical importance (e.g.,
propane liquid is 4.8 times denser than the gaseous
phase).

All of the preceding factors, taken together, determine the time
rate of heat energy release.

RADIATIVE FEEDBACK

All combustion processes, be they of the flaming or of the glowing type, are subject to the stimulating effect of radiative feedback. The larger the fire, the greater is the time rate of heat
energy release because of the greater unit fuel consumption;
hence, the dominance of radiative feedback becomes apparent.
It performs literally as a prime mover needed to activate the entire burning process. It is of such importance that it becomes
necessary to quantify it. Until about 30 years ago, it could not
be satisfactorily measured. Lord Kelvin is credited with the
following quotation:

> I often say that when you can measure what you are speak-
> ing about, and express it in numbers, you know something
> about it; but when you cannot express it in numbers, our
> knowledge is of a meager and unsatisfactory kind; it may be
> the beginning of knowledge, but you have scarcely, in your
> thoughts, advanced to the stage of science whatever the
> matter may be.

Burning Rates of Liquid and Gaseous Fuels

After World War II, various research organizations in Great
Britain, the USSR, and the United States investigated radiative
feedback and were able to express it mathematically by the effect it produced. Of particular importance were the studies of
Van Dolah, Burgess, Wolfhard, Strasser, and Grumer of the
Bureau of Mines, U.S. Department of Interior. Their observations concerned the rates of liquid fuel "burn-off" are directly
related to the rate of heat energy released to the surrounding
environment. They experimented with a wide variety of fuels
and, as would be expected, observed a wide range of radiative
feedback. Their observations and conclusions are hereby summarized, in a general way. A more detailed application of the
basic principles will be presented later in the text.

It had been observed that, under closely controlled test conditions, the rate of liquid fuel burn-off could be formulated in terms of a dimensionless parameter. To obtain this most crucial value it was required that the following stipulations be observed:

1. All fires will be conducted in a free outdoor environment.
2. No air movement will be permitted, except that of the inducted air.
3. Air conditions will be more or less standard.
4. Circular, leveled firepans will be used.
5. All pans will be filled brimming full, for each type of liquid fuel tested, at commencement of the fire test.
6. The burnoff will be measured as the distance down from the lip of the pan.

Figure 1.4 is a graphic portrayal of the testing results of many liquid fuels, ranging from cold liquefied flammable gases, such as liquid hydrogen, liquid natural gas, and liquid propane, to many naturally liquid flammable chemicals. Invariably, the characteristic relationship of circular pan diameter versus the burn-off rate is as illustrated in Figure 1.4.

For small circular firepans, up to about 3 in. in diameter, the flames are laminar, resembling the smooth-burning characteristics of a candle flame and are similar to premixed flames. During this phase the burn-off rate is rapidly reduced to about one-fifth of its initial rate. For intermediate-pan diameters, from 3 in. to about 8 in., the flame fluctuates very noticeably, and burn-off is at a low level. Combustion is so unstable that extinguishment can be easily accomplished. When the firepan exceeds 8 in. in diameter, the flames will have passed the transitional stage into a region of turbulence, and a reduction in fluctuation will ensue. The fuel burn-off rate now is increasing. When the firepan diameter exceeds 2.5 ft, the fuel burn-off rate levels. The observers further noted that as pan diameters increased, the burning rate remained constant, and the fire could be stated as having been in a steady-state condition. This situation now makes it possible to conduct meaningful comparative fire-extinguishing tests. The burnoff rate, in Figure 1.4, is expressed as a function of the dimensionless parameter

$$\frac{H_c}{H_v}$$

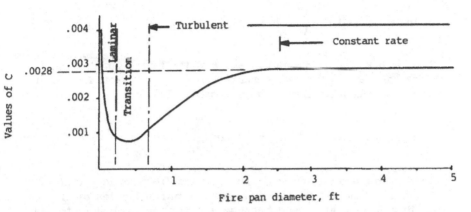

Diffusion Flames Liquid fuel burn-off rate (in./min) = $C(H_C/H_V)$

where

H_c = heat of combustion

H_v = heat of vaporization

C = proportionality constant

FIGURE 1.4 Liquid fuel burn-off rates. Conditions: (1) circular fire pans; (2) brimming full of fuel; (3) outdoor environment; (4) dead calm; (5) standard atmospheric conditions.

where H_c and H_v are, respectively, the unit heat of combustion and the unit heat of vaporization.

It had been anticipated that these two latter factors would be related in their influence on the burn-off rate. It is truly remarkable that such a simple relationship can hold for the wide range of fuels that were tested, various volatilities, some in excess of 1000:1.

Naturally, every flammable liquid has its individual values of H_C and H_V, but their proportion to each other is the determinant, regardless of volatility or viscosity. Because the observers were principally interested in the steady-state conditions, a relationship could now be established:

$$\text{burn-off rate} = c\left(\frac{H_c}{H_v}\right)$$

The ratio H_c/H_v is dimensionless, and the proportionality constant can be determined by measurements of burn-off for given time intervals after fire extinguishment. Hence,

$$\text{burn-off rate (in./min)} = 0.0028\left(\frac{H_c}{H_v}\right)$$

Because burn-off rates were constant for firepans in excess of 2.5 ft in diameter, we can now express burning rates by including firepan area, and because 1 in./ft^2 is equivalent to 144 in.^{3}/ft^2, or 0.62 gal/ft^2, or 5.19(sp gr)lb/ft^2 where sp gr is the specific gravity of a liquid, the total BTU/(min·ft^2) of pan area equals 0.0145(sp gr)$(H_c/H_v)H_c$. Therefore, for a given liquid fuel, the total time rate of heat energy released per minute per square foot of fire area is a constant. The firepan area will now determine the overall time rate of heat energy released per minute for the entire firepan.

Take, for example, the aliphatic hydrocarbon liquid fuel, *n*-heptane (C_7H_{16}),

$$H_c = 20{,}800 \text{ BTU/lb}$$

$$H_v = \quad 138 \text{ BTU/lb}$$

$$\text{sp gr} = 0.8$$

When this compound is burning in firepans exceeding 2.5 ft in diameter

$$\text{burn-off rate} = 0.0028(20{,}800/138)$$
$$= 0.423 \text{ in./min}$$

which is equivalent to

$$0.262 \text{ gal/(min·ft}^2)$$

$$1.74 \text{ lb/(min·ft}^2)$$

and a time rate of heat energy released equal to

$$36,192 \ \text{BTU}/(\text{min} \cdot \text{ft}^2)$$

Note, that in experimental work of this nature, it was necessary for reasons of simplicity to assume complete combustion.
Also, air entry is a linear function of the perimeter of the circular firepan, and the heat-release rate is a function of the
area of the firepan. Hence, as the pans increase in diameter,
the ratio of entering air velocity/weight of fuel consumed increases. Present available fire test data from other sources
show that comparisons between 5-ft^2 and 1200-ft^2 firepans give
no reason to doubt the accuracy of the basic reasoning outlined
so far. Test data are not available to indicate what limitations
exist for larger fire areas.

The preceding example of n-heptane was used because it is
the liquid fuel standard used in the procedure outlined in the
Underwriters' Laboratories, Inc. Standard No. 711 for the testing of hand-portable fire extinguishers. In the conduct of
these tests, certain modifications were required for practical
reasons.

1. Tests were conducted with square firepans.
2. Tests on firepans up to 50 ft^2 were conducted in a special fire-test building with a clear 25-ft ceiling. All
   larger firepans were tested outdoors under no noticeable
   air motion or precipitation.
3. An initial fuel freeboard of 2 in., measured down from
   the lip of the firepan, is necessary to prevent splashing
   from the force of extinguishment.
4. The liquid fuel shall be at a 2-in. depth, floating on an
   8-in. depth of water to prevent heat distortion of the
   firepans.

Heptane was chosen as the standard test fuel because its burning characteristics closely resemble those of gasoline and also
because it is a pure liquid that is not mixed with more or less
volatile, related aliphatic fuels, as is gasoline.

The reproducibility of test data has established this method
as a practical means of rating both hand and wheeled fire extinguishers by matching them in the hands of an experienced
operator. During the past 3 decades that the standard has

Fuel: Heptane

Specific gravity   0.7

Heat of combustion: H_c 20,800 BTU/lb

Heat of vaporization: H_v 138 BTU/lb

H_c/H_v 151

Burn-off rate equivalent to
      0.0028(151) = 0.423 in./min
   or 0.262 gal/(min·ft²)
   or 1.74 lb/(min·ft²)
   or 36,200 BTU/(min·ft²)

Indoor Fires Under 25 Ft
      Test Building
         Ceiling

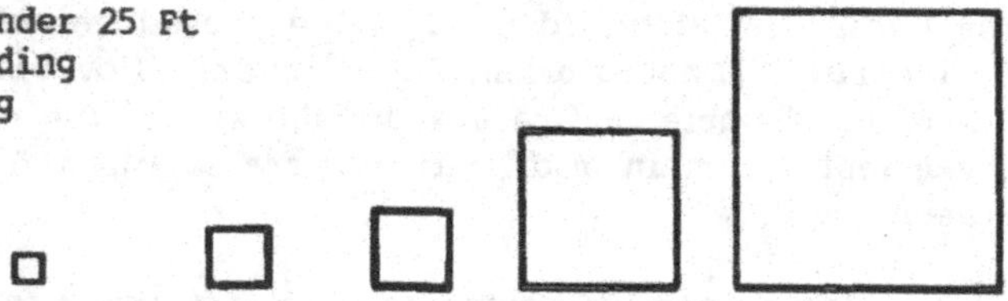

U.L. rating	2B	20B	40B	160B	480B
Area (ft²)	5	50	100	400	1200
Gal/min	1.32	13.2	26.4	105.6	316.8
Lb/min	8.7	87	174	696	2088
*BTU/min	0.181	1.81	3.62	14.48	43.44

*Times 10^6 and assuming complete combustion

FIGURE 1.5 Underwriters' Laboratories test fires (pan) (from
Underwriters' Laboratory Standard 711).

been in force, a unit burn-off rate of approximately 0.42 in./ min has been frequently confirmed within the limits of experimental error. Figure 1.5 shows an array of pans used in the Underwriters' fire tests, from which the total time rate of heat-released energy (BTU/min) can be derived.

Wood Burning Rates

Just as for flaming liquid fuels, radiative feedback plays a vital role for solid fuels. Unfortunately, however, burning rates do not follow the patterns previously described for gaseous and liquid fuels. The burning phenomenon is exceedingly complex and no schema can fit unless the case to be studied is on an individual basis and rigidly adhered to for a particular test condition. No scale model fire tests operating under the same atmospheric conditions as the more expensive full-scale fire tests have ever resulted in any meaningful conclusions. Fire tests for determining the time rate of heat evolution from flaming wood are highly intuitive and are based upon judgmental matters obtained from experience. A review of the *National Fire Protection Association Handbook* on this subject reveals the lack of agreement between data obtained from a variety of test setups. It can be categorically stated that no fire hazard can be fully judged on the basis of any single fire test method. This subsection basically is confined to wood for at least two reasons. First, there is a lot of information about wood and, second, because of wood's vulnerability to fire.

The low heat conductivity of wood is an important physical property. When one considers the situation of heat or flame exposure on a wood surface, charring immediately takes place and proceeds inwardly. The char formation possesses an even lower heat conductivity than does the wood, varying from 40 to 60% as much, depending upon the type of wood and whether heat conduction is across or along the grain. Because of low heat conductivity, the surface temperature increases rapidly, as contrasted with a pool of burning heptane, for which immediately after extinguishment the liquid is cool to the touch. The heat retentivity of wood is of considerable importance.

It is difficult to separate and assess the variables involved in the penetration of wood by high temperatures such that ignition and combustion can occur. As stated earlier, if they are to burn, wood masses must be heated to a point at which combustible gases are evolved at the wood's surface. Gases of this type require ignition sources of from 800 to 1000°F before combustion can occur. Open flames of any type have temperatures in excess of 2000°F. As a consequence, it is a relatively simple matter to describe the ignition point of wood as being in the vicinity of 350 to 400°F if an open flame is present to act as an ignition point for the gases evolving from the wood at the ranges of temperature. The density, state of subdivision, amount of pitch or resin content, state of dehydration, and the time rate of heating of the wood, all influence the ignition temperature.

To emphasize the points brought out in the foregoing paragraph, reference is made to research recorded at the Forest Products Laboratory of the US Department of Agriculture. Their findings, represented in Figure 1.6 show the length of time before wood specimens, maintained at the specified constant temperatures, evolved combustible gases in quantities sufficient to be ignited by a pilot flame located about 1/2 in. above the specimen that was located vertically in a 3-in. diameter by 10-in. long vertical quartz chamber. The chamber was maintained at a constant temperature by an electric furnace. The wood specimens were oven-dried. It is obvious from Figure 1.6 that ignition temperatures are greatly affected by time of exposure to a constantly maintained temperature. For the spontaneous ignition temperature of wood, even more complexities occur. The influence of the variables involved is even more indeterminate than when an igniting flame for the evolving gases is present. Considering that wood slowly changes its chemical composition under sustained heat attack, the endpoint of such an exposure, over long periods, is the formation of extremely porous charcoal that has notable catalytic properties. The data are virtually meaningless without a time frame. Wood in contact with steam pipes or a similar constant-temperature source over a long period undergoes chemical changes that result in the formation of charcoal, which is capable of heating spontaneously. The only conclusions that can be drawn from all this relate to prevention of wood decomposition if self-ignition is not to occur. The various building codes try to cover this by limiting full-time exposure to heat sources with ambient temperatures not in excess of 150°F to eliminate the risk of spontaneous ignition.

The variables inherent in any test procedure for estimating heat release rates and extinguisher capabilities would include

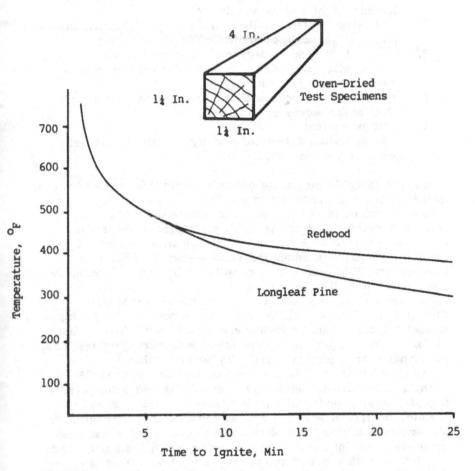

FIGURE 1.6 Time required to ignite wood specimens using pilot ignition.

1. *Size and shape*: Ratio of exposed surface to bulk,
    varying from wood shavings as one extreme, to strips,
    planking, light and heavy timber.
2. *Positioning*: Effect on fire behavior by horizontal, in-
    clined, or vertical positioning of wood elements.
3. *Array*: Influence on degree of aeration and radiation
    such as with loose piles, compact piles, and various lat-
    ticed crib arrangements.
4. *Environment*: Indoor and outdoor locations and their
    effect on air movement.
5. *Moisture content*.
6. *Type of wood*: Effects of density and the content of
    gums, resins, and oils.

To make intelligible judgments possible among this tangle of variables, it becomes mandatory to confine the fire testing to particular situations involving limits that embrace all of the variables just listed. The key is to obtain reproducible test data
and to obtain thereby a socalled yardstick in comparing materials and extinguishers under the same operating conditions.
Furthermore, this cannot be accomplished by trying to compare
small-scale and full-scale test data.

So, once again, we refer to Underwriters' Laboratories,
Standard No. 711, wherein is outlined a procedure for testing
manually operated fire extinguishers against wood fires. Just
as for liquid fuel pan fires there are a wide variety of test fire
sizes that bear a geometric similarity to each other.

As an example, we will look particularly at the class 2A test
such as illustrated in Figure 1.7. When this test setup was being prepared, a review of burning tests of various-sized cribs,
mounted on special weighing scales, revealed that although overall constant-burning rates were not attainable, there was, nevertheless, a definite time interval during which the burning rate
was indeed both constant and at its peak value. The plot of heat
flux versus time, shown in the lower part of Figure 1.7, is un-
derstandably an idealized form. As shown in this figure, the
wood elements that form the crib are defined for type, size,
and moisture content. Furthermore, the crib is positioned a
fixed distance above the floor of the fire-test building. Initially 2 qt of n-heptane serves as the means of ignition. By the
end of 4 min, the liquid fuel will have completely burned away
and, from previous test findings, approximately 10% of the ini-

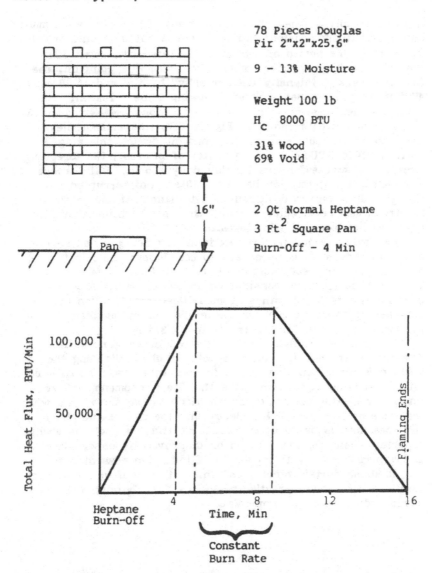

78 Pieces Douglas
Fir 2"x2"x25.6"

9 - 13% Moisture

Weight 100 lb
H_c 8000 BTU

31% Wood
69% Void

2 Qt Normal Heptane
3 Ft^2 Square Pan
Burn-Off - 4 Min

16"

Pan

Total Heat Flux, BTU/Min

100,000

50,000

Heptane
Burn-Off

Time, Min

Constant
Burn Rate

Flaming Ends

4   8   12   16

FIGURE 1.7 Underwriters' Laboratories test fires (crib) (from
Underwriters' Laboratory Standard 711).

tial crib weight will have been consumed. By 5 min, a maximum
and substantially constant burning rate is obtained until minute
9. During this period of maximum fire intensity, about 60% of
the crib weight is further consumed. From this point on, the
fire decreases in intensity and, usually, by the end of 16 min,
all flaming has stopped and only glowing embers remain. When
extinguishment testing is performed it is done between 7 and 8
min. From the data shown in Figure 1.7, the maximum heat
flow rate to the surrounding environment during this stage is
about 120,000 BTU/min. As a matter of interest, the foregoing
fire can be successfully extinguished by a 5-lb, multipurpose,
dry-chemical extinguisher having a 10-sec uninterrupted discharge. This test is designed to determine that the extinguisher stream can be directed through the maze of internal hidden
openings between the crib elements.

Now, to determine that the extinguisher being tested also has
adequate ranges, it is necessary to complement the crib fire test
with another fire test using a 10 ft. × 10 ft. vertical panel.
This panel is specially constructed to be self-sustaining. The
burning side is faced with a cross-latticed construction of furring having the same type and moisture content as that used in
the crib test. The furring is 3/4 in. × 3/4 in.

The first half of this chapter covered broad generalities.
The remainder went into considerable detail highlighting the
various factors, how they relate to each other, and how their
interdependence combine to make the fire phenomenon either
terrifying as the master, or subservient as the slave. As with
the state of our general knowledge and for fire, in particular,
our reasoning is based upon empiricism and, as such, is always
capable of being either verified or disproved by observation
and experiment. By this process of inductive reasoning we
can establish certain tenets, call this a theory if you will, that
combined with sound practice provide a firm foundation in our
never-ending quest for improvement.

2
Flammability Characteristics of Combustible Gases and Vapors

The prevention of unwanted fires and gas explosion disasters requires a knowledge of the flammability characteristics of combustible gases and vapors likely to be encountered under various conditions of use (or misuse). Available data may not always be adequate for use in a particular application because they may have been obtained at a lower temperature and pressure than is encountered in practice. For example, the quantity of air required to decrease the combustible vapor concentration to a safe level, in a particular process carried out at 400°F, should be based on flammability data obtained at this temperature. When these are not available, suitable approximations can be made to permit a realistic hazard evaluation associated with the process being considered. Such approximations can serve as a basis for designing suitable safety devices for the protection of personnel and equipment.

An understanding of these characteristics is vital to the proper operation, inspection, and maintenance of the equipment. Fire protection, either by automatic or manual means, as well as lowering the level of hazard to the lowest practical extent, also requires a knowledge of these characteristics.

FLAMMABILITY LIMITS

A combustible gas—air mixture can be burned over a wide range
of concentrations, either when subjected to elevated temperatures or exposed to a catalytic surface at ordinary temperatures.
However, homogeneous combustion gas—air mixtures are flammable, that is, they can propagate flame freely within a limited
range of compositions. For example, trace amounts of methane
in air can be readily oxidized on a heated surface, but a flame
will propagate from an ignition source at ambient temperatures
and pressures only if the surrounding mixture contains at least
5% but less than 15% methane by volume. The more dilute mixture is known as the lower flammability limit (LFL) or the com-
bustible-lean limit mixture; the more concentrated mixture, is
known as the upper flammability limit (UFL), or combustible-
rich limit mixture. Refer to Figure 2.1 and note that the area
of the plot between the upper and lower limits comprises vapor
concentrations that are flammable. The horizontal line, termed
"stoichiometric mixture," represents that vapor concentration in
air that is neither rich nor lean and has the theoretical amounts
of vapor and air needed for complete combustion. The velocity
of flame propagation is at a maximum value at this latter ideal
concentration. Velocity of flame propagation tapers off to zero,
technically speaking, at either limit. Also, note that the limits
of flammability diverge as the temperature rises. Figure 2.2
illustrates this divergency in detail for methane in air at atmospheric pressure. The slope of the limit lines are expressed as
the ratios of the respective limits in terms of the limits under
standard conditions versus temperatures. Methane was chosen
as our example because there is ample data. Other paraffinic
hydrocarbons show similar behavior. Unfortunately, data are
sparse for other hydrocarbons. The data were compiled in U.S.
Bureau of Mines Bulletin 627. Assuming that methane has values of 15 and 5%, respectively, for the upper and lower limits
of flammability and at a reference temperature of 70°F (21°C)
then

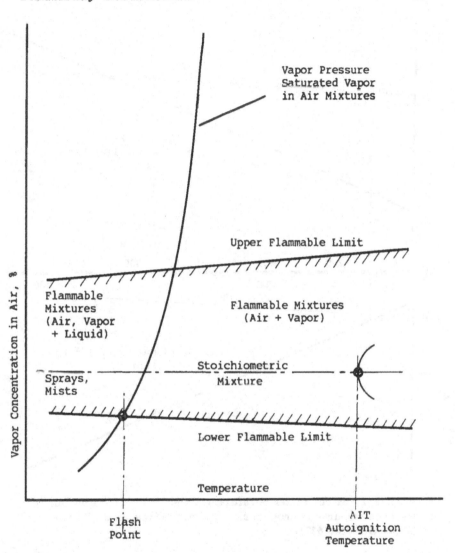

FIGURE 2.1 Basic flammability factors.

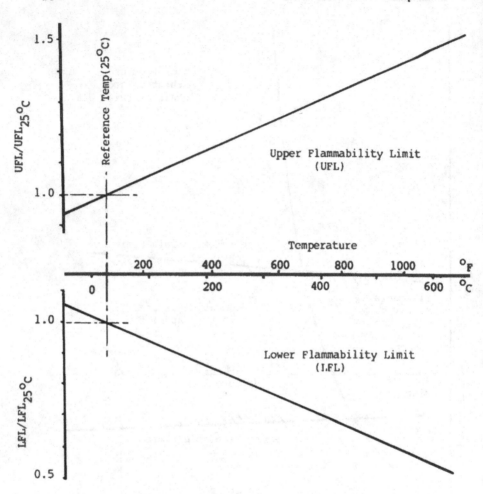

FIGURE 2.2 Effect of temperature on the limits of flammability of methane in air at atmospheric pressure (from U.S. Bureau of Mines Bulletin 627).

	Temperature	
	70°F (21°C)	1000°F (538°C)
UFL (%)	15	21
LFL (%)	5	3

Hence, there is an 80% increase in the flammability range at 1000°F over that at 70°F.

VAPOR PRESSURE

The general properties of liquids have been accounted for by the kinetic theory, examples of which are

1. Liquids are practically incompressible. Unlike gases, but like solids, there is a microscopic change in liquid volume when the pressure is raised, even up to thousands of atmospheres. The kinetic theory accounts for this by saying that the amount of free space between molecules of a liquid has been practically reduced to a minimum. Any attempt to compress the liquid further meets with resistance as the electron cloud of one molecule repels the electron cloud of an adjacent molecule.

2. Liquids maintain their volume regardless of the shape or size of the container. Gases do not conserve their volume because the molecules are essentially independent of each other and move into any space that is available. In liquids the molecules are very close together and mutual molecular attractions are strong.

3. Because liquids have no characteristic shape, they assume the shape of the containers in which they reside. The kinetic theory explains this property by saying that there are no fixed positions for the molecules, which is what distinguishes liquids from solids. The molecules are free to move and slide over each other to occupy positions of the lowest possible potential energy. On earth, gravity pulls the liquid specimen to the bottom of its container; in an orbiting satellite, intermolecular forces pull the specimen into a spherical glob.

4. Liquids diffuse slowly. When a drop of ink is carefully released onto water, there is at first a sharp boundary between

the ink cloud and the water. Eventually the color diffuses
throughout the rest of the liquid. In gases, diffusion is much
more rapid. Diffusion occurs because molecules have kinetic
energy and move from place to place. In a liquid, molecules do
not move far before colliding with neighboring molecules. The
mean free path, which is the average distance between collision,
is short. Eventually, each molecule of a liquid does migrate
from one side of the container to the other, but it suffers bil-
lions of collisions in doing so. In gases there is far less ob-
struction to the migrating molecule. Because a gas is mostly
empty space, the mean free path is much longer. Hence, molecules of one gas can quickly mix with others, i.e., diffuse more
quickly.

    5. Liquids evaporate from open containers. Although there
are attractive forces that tend to hold the molecules together,
it is evidently possible for molecules to escape. Those molecules with kinetic energy great enough to overcome the attractive forces can escape into the gas phase. In any collection, a
given molecule does not always have the same energy. There
is a perpetual interchange of energy on collision. If we sim-
plistically assumed that we were to start with all molecules of
the same energy and velocity, we would quickly perceive that
this condition cannot exist. Two or more molecules may simultaneously collide with a third. Molecule 3 now has not only its
original energy but possibly some extra energy received from
its neighbors; it now has higher-than-average kinetic energy.
If it happens to be near the liquid surface, it may be able to
overcome the attractive forces of its neighbors and exit off into
the gas phase. As these hyperenergetic molecules leave the
liquid phase, the average kinetic energy of those left behind is
lower, because each molecule that escapes carries along with it
more than an average amount of energy, part of which is used
in working against the attractive forces. Because the average
kinetic energy of the molecules remaining in the liquid phase is
lower, their temperature drops; therefore, evaporation is accompanied by cooling.

    If now the liquid is not free to evaporate from an open con-
tainer (say, by means of a lid or cover) hyperenergetic molecules will continue to escape from the liquid phase. However,
after escaping they are confined to a limited space. As the
molecules accumulate in the space above the liquid, there is an
increasing tendency that, as a result of their random motion,
some of them return to the liquid. Eventually, a situation is

established in which the molecules return to the liquid as fast
as others are leaving it. This latter condition is referred to as
"dynamic equilibrium." Although the system is not at a state of
rest, there is no net change in the system. The molecules that
are in the vapor phase exert an independent pressure which, at
equilibrium, is characteristic of the specific liquid at a given
temperature and is known as the "equilibrium vapor pressure."
As the term implies, it is the pressure exerted by a vapor when
in equilibrium with its liquid. The magnitude of this pressure
depends only on

1. The nature of the liquid
2. The ambient temperature

There are various devices for measuring equilibrium vapor
pressure, but a simple and practical one is illustrated in Figure
2.3. It consists of a mercury atmospheric barometer set up in
the usual manner. The difference between the upper mercury
level and the lower level represents the atmospheric pressure.
Above the mercury there is a vacuum. By squeezing the medicine dropper, a couple of drops of liquid (to be tested) is injected as shown. Because practically all liquids are less dense
than mercury, they float to the top of the mercury where enough
liquid evaporates to establish its equilibrium vapor pressure under the prevailing ambient temperature condition. (When making
this determination, be certain that a slight excess of test fluid
floats on top of the mercury column.) The vapor pressure
pushes down the mercury column. The extent to which the mercury level is depressed gives a quantitative measure of the vapor pressure of the test liquid.

The nature of the liquid is involved because each liquid has
characteristic attractive forces between its molecules. Molecules
that have a large mutual attraction have a small tendency to escape into the vapor phase, resulting in a low-equilibrium vapor
pressure. Liquids composed of molecules with small mutual attraction have a high escaping tendency and, therefore, a high-
equilibrium vapor pressure. In so far as temperature is concerned, a temperature rise means that the average kinetic energy of the molecules is increased. The number of high-energy
molecules capable of escaping becomes larger; hence, the equilibrium vapor pressure increases.

Over 100 years ago, the French physicist Trouton observed
that the heat of vaporization (H_V) is approximately related to

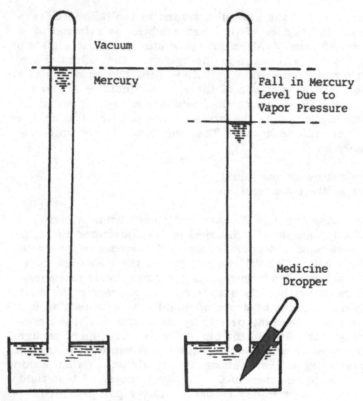

FIGURE 2.3 Measurement of vapor pressure.

the absolute temperature at the boiling point. Because the heat of vaporization is a measure of the energy required to separate the molecules of the liquid, i.e., to overcome the intermolecular forces, the boiling point serves as a convenient index of the strength of these forces. Figure 2.4 illustrates the relationship of vapor pressure with temperature for several of the lighter liquid paraffinic hydrocarbons that are the main components of gasoline.

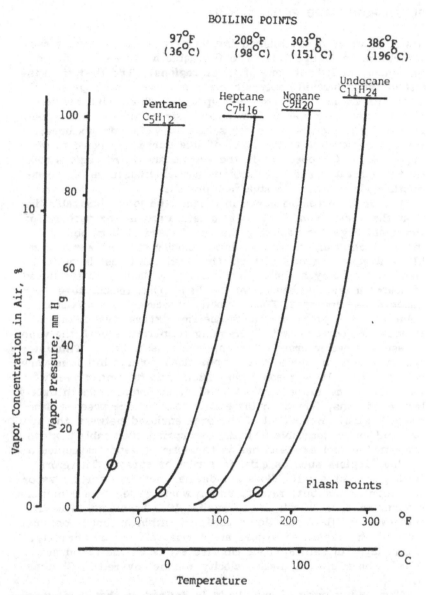

FIGURE 2.4 Effect of temperature on vapor pressure of several lighter paraffinic hydrocarbons.

FLASH POINT AND BOILING POINT

Now that we have established the vapor pressure curve, please
refer back to Figure 2.1. The flammable mixtures considered
in Figure 2.1 fall into one of three regions. The first is to the
left of the saturated vapor-air mixture curve (vapor pressure)
in the region labeled "mists and sprays." Such mixtures consist of droplets suspended in a vapor-air mixture. The second
region lies along the curve for saturated vapor-air mixtures;
the last region lies to the right of this curve and consists of
vapor only. Compositions in the second and third regions make
up the saturated and unsaturated flammable mixtures of a com-
bustible-air system at a specified pressure.

The vapor pressure curve intersects the lower flammable limit at the "flash point," which is a term used in the parlance of
fire technology for depicting the fire hazard of flammable liquids. Technically speaking, flame velocity at the lower flammable limit equals zero, and when the flash point test is performed
there will be only a visible, nonsustaining flash until the liquid
is heated a few degrees above the flash point temperature. As
temperatures increase, flame velocity increases to a maximum
value at the approximate stoichiometric mixture level from,
whence, it decreases with increasing temperature until the vapor
pressure curve intersects the upper flammable limit. This inter-
section has been labeled the "upper flash point," but it is very
seldom used. It represents zero flame velocity under overly
rich mixture conditions. Most studies, however, concern overly
lean conditions, for which increasing temperature poses a flammable hazard. Hence, all of the area enclosed between the upper and lower flammable limit lines comprises flammable mixtures
wherein the fuel exists either in the form of vapor molecules or
of fine droplets such as exist in a mist or spray. Therefore,
flash point, as a term, does not define whether a certain vapor
will burn or not but, rather, under what temperature conditions
combustion will, or will not, prevail. It is important to realize
that even for flammable liquid confined within an open container
from which evaporated vapors are carried off by air currents,
a thin layer of saturated air mixture exists on the liquid surface. The subject of flame velocity will be covered in Chapter
4.

The boiling point of any liquid is defined as that point during heating when the vapor pressure becomes equal to the external pressure exerted upon the liquid. At that point all intermolecular attraction ceases.

FLAMMABLE LIQUID CLASSIFICATION SYSTEMS

For fire protection purposes, an arbitrary division of liquids
and gases has been based upon the definition of a flammable
liquid that appears in NFPA Standard 321, *Basic Classification
of Flammable and Combustible Liquids*. Liquids are those fluids
having a vapor pressure not exceeding 40 psi absolute pressure
at 100°F (38°C). Furthermore, this NFPA Standard, which is
a part of the National Fire Codes, is based upon a concept illustrated in Figure 2.5, wherein liquids that will burn are divided into three categories.

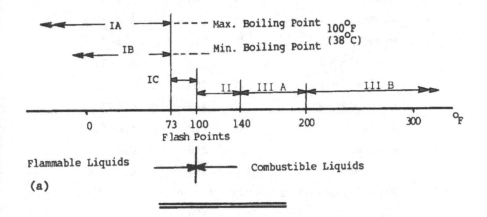

FIGURE 2.5 Classification systems of flammable and combustible
liquids: (a) National Fire Code classification; (b) Underwriters'
Laboratories classification (from National Fire Protection Association Standard 321).

TABLE 2.1 Subclassifications of Some Flammable Liquids

Class	Liquid	Flash point (°F)	Boil. point (°F)	Range of vapor flammability
IA	Ethyl ether	−49	90	1.9−36.0
	Pentane	−40	99	1.5−7.8
	Acetaldehyde	−36	70	4.0−60.0
IB	Carbon disulfide	−22	115	1.3−50.0
	Hexane	− 7	156	1.1−7.5
	Acetone	0	134	2.6−12.8
	Benzol	12	176	1.3−7.1
	Ethyl alcohol	70	173	3.3−14.0
IC	Amyl acetate	77	300	1.1−7.5
	Butyl alcohol	84	243	1.4−11.2
	Nonane	88	303	0.8−5.4
II	Decane	115	345	0.8−5.4
IIIA	Undecane	149	384	0.7−5.4
IIIB	Paraffinic (lub.) oil	(300)	Above	
		(450)	650	

1. It is anticipated that in most areas the indoor temperature can reach 100°F (38°C) at sometime during the course of a year. Therefore, all liquids with flash points below 100°F are called class I liquids.
2. In some areas the ambient temperature could exceed 100°F, at which only a moderate degree of heating would be required to heat the liquid to its flash point. Given this concept, an arbitrary division of 100°F to 140°F (60°C) was established for liquids with flash points within this range, that would be known as class II liquids.
3. Because liquids with flash points higher than 140°F would require considerable heating from a source other than ambient temperatures before ignition could occur, they have been classified as class III liquids.

TABLE 2.2 Underwriters' Laboratory Classifications of Some Flammable Liquids

Relative flammability hazard	Common examples
100	All class IA liquids
90–100	Hexane, acetone, benzol
60–70	Ethyl alcohol, Amyl acetate
30–40	Butyl alcohol, nonane, decane, undecane
10–20	Paraffin (lube) oils

Class I and III liquids are each divided into subclasses that are shown in Figure 2.5. Examples of various liquids are listed in Table 2.1.

Note that the classification system is based upon flash points. Class IA liquids are further distinguished by having boiling points not exceeding 100°F. Class IB liquids have boiling points in excess of 100°F. Additionally, note that boiling points and ranges of flammability bear no real relationship to flash points in this system of classification. As important as these factors are in the overall picture, it cannot be denied that flash points indicate the ease of ignitability, and that is all. The flash point has no relationship to the degree of fire intensity. Figure 2.5 also shows another system of evaluating flammability hazards. For many years the Underwriters' Laboratories has had a system for grading the flammability hazards of various liquids. The classification is based upon the following scale shown in Table 2.2. The same examples of liquids used in describing the NFPA Standard 321 classification are also tabulated. The reason for showing these two classification is to illustrate the similarity of reasoning. In both tables, the flash point is considered an index for ease of flammability.

THE IGNITION PROCESS

Autoignition

When a flammable mixture is heated to an elevated temperature, a reaction will ultimately be initiated that will proceed with sufficient rapidity to ignite the mixture without the use of any separate ignition means. The time that elapses between the instant the mixture temperature is raised and that when a flame appears is loosely referred to as the time lag or time delay before ignition. Two styles of autoignition data are found in the current literature.

In the first type, the effect of temperature on time delay is considered for delays for less than 1 sec. Such data are applicable to systems in which the contact time between the heated surface and a flowing flammable mixture is very short, but these data are not satisfactory when time is indefinite. Hot surface ignition is an uncertain matter because with volatile flammable liquids (class I; see Fig. 2.5), immediate vaporization results and the heated surface is enveloped with an overly rich vapor-air mixture unless ventilation comes into action. Class III combustible liquids are easily inflamed and class II liquids occupy an intermediate position.

For the second type, it is the lowest temperature at which ignition can occur that is of interest. This temperature is called the "minimum spontaneous ignition" or the "autoignition" temperature and is referred to as AIT (see Fig. 2.6). From the safety standpoint, the AIT is universally accepted as the maximum temperature to which a flammable mixture can be heated without flaming. From Figure 2.1, it will be seen that the AIT is a minimum for the stoichiometric fuel/air ratio wherein the amounts of air and flammable vapor are theoretically exact relative to each other, there being no excess of one over the other. The tests to determine the AIT are carried out in a large uniformly heated pressure vessel under closely controlled conditions. As would be expected from the kinetic theory of gases, an increase in pressure generally decreases the AIT of a combustible or flammable substance mixed with air. For example, the US Bureau of Mines found that the AIT of methane in air decreased from 986°F (530°C), at 1 atm to 464°F (240°C) at 610 atm. Injection of a liquid fuel into a heated container at temperatures above 1100°F (594°C) will explode in a very short time (in the order of milliseconds). A common useful example of the forego-

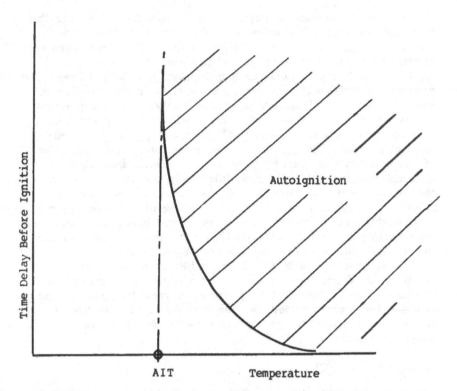

FIGURE 2.6 Determination of minimum autoignition temperature (AIT).

ing is the diesel engine. An uncommon (and fortunately so) example is the malfunctioning multistage air compressor, in which because of overheating, the lubricating oil exploded.

Ignition Energy

The ignition process is initiated by the application of energy.

1. In the form of heat manifesting itself by an elevation of temperature.
2. In the form of actinic electromagnetic radiation with the total effect being independent of temperature.

In both forms, chemical changes are produced by energizing the outermost electrons of a molecule. The extent of energizing varies with the energy level applied, which varies from mere activation to a complete severance of the electron leaving a positively charged free radical (or a molecular fragment). Free radicals are extremely reactive because of their tendency to regain an electron and restore electrical balance. The deprivation of the electron(s) results in the free radical seeking another reactant to form a new product of lower energy level and releasing free heat energy that serves to continue the stimulation of the reaction (exothermic reaction). A classic example of a chemical reaction initiated by actinic electromagnetic radiation is the direct union of hydrogen and chlorine. We are used to thinking of oxygen as the sole oxidizing agent. There are other oxidizing agents as well, a prominent one being chlorine. An oxidizing agent is a material that can oxidize a fuel (reducing agent) and, by doing so, becomes reduced itself.

The process is one in which the oxidizing agent acquires electrons by depriving the fuel of its electrons. When chlorine is mixed on a 1:1 basis with hydrogen at room temperature, no reaction takes place as long as the mixture is kept in the dark. In diffused sunlight the two gases combine slowly to form hydrogen chloride:

$$H_2 + Cl_2 \rightarrow 2HCl$$

In direct sunlight, or in the light of a burning magnesium ribbon or the light from an electric arc, the reaction is instantaneous—an explosion! The activation of this reaction (ignition) is caused by the presence of ultraviolet radiation of 0.3 μm (1 μm = 10^{-6} m) wavelength. In the preceding paragraph, it was stated that the ignition process was initiated in either of two ways.

1. By supplying heat energy that imparts a more rapid and chaotic movement to the molecules affected and manifests itself as an increase in temperature. As the heat energy is increasingly applied, the molecules acquire an increase in their kinetic energies, as evidenced by their individually higher speeds, and they collide ever more forcefully with each other. Finally, when enough energy has been applied to emit free electrons and the resulting free radicals are released, then and there, a chemical reaction starts. The various bonds that join the atoms to

the molecules have varying dissociation energies. As is to be
expected, the weakest bonds will break first. Energy in the
form of heat will now be released with the formation of new molecules, and the overall reaction will proceed to completion.

2. By supplying electromagnetic radiation energy to a molecule, vibrations are set up at the various electronic-bonding
junctions within the molecule. Because the resistance to the
breaking of these electronic-bonding points varies, it is evident
that the weakest ones will be the most vulnerable. If the electromagnetic radiation is sufficiently actinic, the weakest bond
will be subjected to resonance, and the same final effect of
electron emission, formation of free radicals, emission of heat
energy, and the formation of a new molecule, will result.

As is true for all energy transfer, there is the element of
time. When a flammable gas or vapor is subjected to high temperatures for a very short interval there may be no ignition.
On the other hand, ignition may occur when the flammable substance is exposed to lower temperatures for a longer period.
This behavior is particularly noticeable in autoignition. Early
combustion reactions following ignition can be classified as either

1. Uniform reactions, wherein the reaction occurs throughout the entire mixture at substantially the same time.
   This is what occurs with autoignition at lower temperature levels. At higher temperature (more energy input),
   as would be expected with pilot or spark ignition, the
   shock effect of the faster-occurring reaction becomes increasingly explosive.
2. Propagating reactions in which there is a clearly defined
   reaction zone which radiates outwardly from the source
   of ignition through the unreacted materials.

Once a flammable mixture is ignited by a source of ignition,
the flame will either attach itself to the ignition source, as with
a pilot gas flame, or propagate away from it, as with electric
ignition. If the flame propagates away from the ignition source,
the rate of reaction is either subsonic (deflagration) or supersonic (detonation). Further studies of these phenomena are
possible by referring to Bulletin 627 of the US Bureau of Mines
authored by M. Zabetakis.

Let us look deeper into the subject of ignition and first consider autoignition as was described earlier in this chapter. Fig-

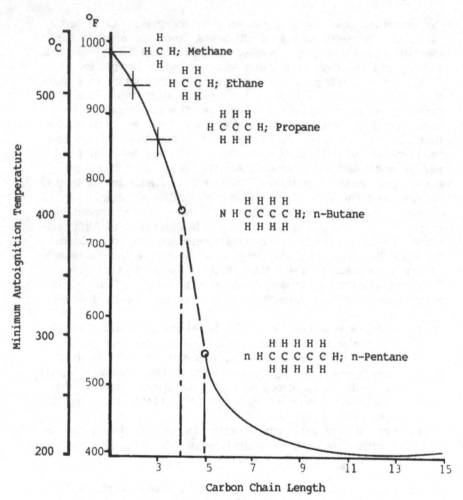

FIGURE 2.7 Minimum autoignition temperatures of normal paraffinic hydrocarbons (C_nH_{2n+2}; n = total number of carbon atoms) in air as a function of carbon chain length.

ure 2.7 illustrates certain examples for which considerable data
are available because they are closely related to the field of
petroleum chemistry. Note, that the AIT of the normal paraffinic hydrocarbons are plotted versus the carbon chain length,
i.e., the number of carbon atoms forming the so-called spine
of the molecule. It is most obvious that, as the carbon chain
length increases, the normal molecular structure becomes more
vulnerable to rupture by autoignition temperature. Also note,
that a discontinuity exists in the curve between 750°F (400°C)
and 540°F (282°C) and between normal butane and normal pentane. The upper portion of the curve commences with methane
gas, which has an AIT of nearly 1000°F (532°C). With each
additional atom of carbon, the AIT decreases by an increasing
amount until normal butane is arrived at (carbon chain length =
4). The addition of one more carbon to the chain length results in a sharp drop in the AIT. From that point on the drop
in AIT with increasing carbon chain length becomes less noticeable, leveling out at 400°F (204°C). Once more, Bulletin 627
of the US Bureau of Mines by M. Zabetakis should be referred
to.

Thus far, discussion has been limited to "normal" molecular
structures in which the carbon atom chain was substantially
straight. Figure 2.8 illustrates the molecular structure of normal pentane, as just described. It is, however, possible to
chemically rearrange the five carbon and 12 hydrogen atoms in
two different manners because of the different valences of the
carbon atom (+ or − 1) and the hydrogen atom (+ 1). Such
rearrangements result in molecules that are called "isomers,"
which here would be methyl butane and dimethyl propane. Note
the difference in their properties: as the number of straight-
line carbon atoms is reduced the liquid pentane becomes increasingly more volatile (lower boiling point) and also becomes more
resistant to autoignition (higher AIT). Obviously, as the molecule becomes more contracted and acquires, as a result, more
methyl radicals (CH_3), greater thermal stability is achieved.

A practical example of this process of "isomerization" has
been the partial conversion of normal paraffinic hydrocarbons,
such as pentane, hexane, heptane, octane, and nonane, to
some of their isomers in the manufacture of unleaded gasoline.
This results in the general elevation of the AIT levels and re-
tards any tendency to detonate or "ping" in gasoline engines.
The extent to which this is accomplished is considerable when

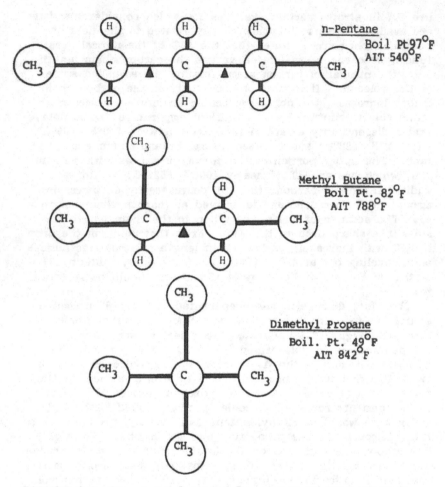

FIGURE 2.8 Structural formulae for normal pentane (C_5H_{12}) and its effects on volatility (AIT). Arrowheads indicate bonds most vulnerable to ignition energy.

one takes into account that as the pressure is raised (compression stroke), the AIT becomes lower in value unless the isomeric constituents are used to overcome this effect.

There are other means of raising the AIT of liquid fuels, including the addition of alcohols or ring compounds, such as benzol and toluol, all of which have relatively high AIT values. These procedures were developed largely from the attempt to eliminate the use of tetraethyl lead because of toxic considerations. This latter compound was added in small amounts (up to about 4 cm^3/gal) because its thermal decomposition released lead atoms which served as inhibitors in moderating the rate of cylinder pressure rise.

Previously, it was stated that ignition would be initiated by either temperature or actinic electromagnetic radiation when they were at, or above, energy levels to provide sufficient molecular motivation. It must be emphasized that these twin aspects of ignition are not mutually exclusive but are closely related to each other. About 100 years ago, after the establishment of the kinetic theory of gases, four physicists made history in their study of molecular energy and its implications. Their findings completely confirmed each other.

1. *Boltzmann* established the fundamental law by which the kinetic energy per molecule is related to the absolute temperature.
2. *Stefan* established the basic theory of "black body" heat radiation and its relation to the fourth power of its absolute temperature.
3. *Wien* related the wave length of emitted radiation to that part of the spectrum at which the temperature of a given flammable substance was at a maximum.
4. *Planck* developed an empirical equation that satisfactorily represented the observed energy distribution in the spectrum of a black body, thus confirming by his quantum theory the observations of the earlier three investigators.

The reader is referred to basic texts on these subjects because it is not feasible to expand further within the confines of this book.

Figure 2.9 is based upon Boltzmann's law which expresses the effect of temperature upon molecular energy (all of the

Average kinetic energy per molecule = 3/2KT
where K = Boltzmann constant = 8.3 × 10^{-5} eV/molecule/°K
and T = temperature (°K)

Hence $K.E._{av}$ = 12.5 × 10^{-5} eV/°K

Condition and Fuel Types	Temperature			Range of Max. Energy Available for Ignition (eV)	
	°F	°C	°K	From	To
Autoignition					
Carbon disulfide	81	27	300	0.038	0.114 - 0.142
Aldehydes					
Ethers	261	127	400	0.050	0.150 - 0.190
Fuel oils					
Glycols	441	227	500	0.062	0.186 - 0.238
Alcohols					
Ketones	621	327	600	0.075	0.225 - 0.285
Ring compounds					
Natural gas	801	427	700	0.088	0.264 - 0.333
Hydrogen					
Carbon monoxide					
	981	527	800	0.100	0.300 - 0.380
	1161	627	900	0.112	0.336 - 0.428
Gas pilot flame	2780	1527	1800	0.224	0.672 - 0.840
Oxy-hydrogen flame	4400	2427	2700	0.336	1.008 - 1.260
Oxyacetylene flame	6020	3327	3600	0.448	1.344 - 1.680
High energy spark (welding of steel)	9500	5260	5533	0.700	2.100 - 2.620

FIGURE 2.9 Effect of temperature on molecular energy (eV)
with various types of ignition.

units are indicated at the top of the figure). The right-hand vertical column has two values, in each horizontal row, that confine the probable limits of the range of maximum energy (measured in electron volts) available for ignition. They are based upon a probability variance from the average energy level derived from the Boltzmann equation. The first seven horizontal rows are all concerned with autoignition and are arranged in ascending gradations of temperature, starting from room temperature conditions to a maximum value of 900°K, which is about as high as any autoignition temperatures reported in various references. Carbon disulfide (CS_2) has one of the lowest AIT levels at 350°K. On the other hand, carbon monoxide, has one of the highest values, 900°K. The next four horizontal rows all involve separate methods of ignition that vary from a simple gas pilot flame to a high-energy electric spark.

As practical examples of autoignition one can cite

1. Overheated electrical wires and equipment.
2. Overheated bearings caused by friction.
3. Overloaded or underinflated rubber tires.
4. Spontaneous ignition: shielded decomposition heating in cotton bales, coal piles, grain silos, and organic refuse. Reaction times are usually slow.
5. Automatic instantaneous ignition as a result of otherwise chemical reactions, for instance, a strongly oxidizing pool water disinfectant (calcium hypochlorite) coming into contact with gasoline. There are many more examples.

As practical examples of pilot and spark ignition one can cite

1. Open flames, candles, matches, gas ranges, water heaters, leaking exhaust manifolds, and the like.
2. Incandescent sparks during use of grinding wheels.
3. Glowing drops of molten metal during the use of welding and cutting torches.
4. Incandescent surfaces, such as with a shattered incandescent light bulb that was lit when shattered.
5. Electric sparks from static electricity and also from induction of high-frequency electromagnetic waves in open circuits. A typical example involves minimum spacing between radar equipment and fuel-servicing apparatus at airports (refer to NFPA Standard 407).

6. Sudden conversions of kinetic energy in moving bodies
to heat energy, such as shock or impact to containers
of flammable substances.

It will be noticed that the height to which the temperature
rises is directly related to the molecular energy expressed in
electron volts. The temperature existing within an electric
spark has been estimated, by photoelectric means, to be in excess of 5800°K (about 10,000°F; 5540°C). In general, many
flammable mixtures can be ignited by sparks having a relatively
small energy content of from 1 to 0.1 J (1 J = 6 × 10^{18} eV).
However, the ability of an electric spark to ignite flammable
vapor-air mixtures depends upon the toal energy of the spark
and the time lapse involved in the spark's duration, which is
related to the best dissipation characteristics of the energy
involved. Therefore, although the total energy seems small,
the spark has a large power density in excess of 1 kw/mm^3.
This huge power density, together with a minute amount of
matter existing in the path of the spark, account for the extreme temperatures. For further study, the reader is referred
to NFPA Standard 493, which provides information for the design of intrinsically safe exposed electrical equipment and the
limitation of energy that can be used in class I hazardous locations as defined in the National Electrical Code. The curve
shown in Figure 2.10 portrays the energy spectrum for the
specific temperature of 2371°K (3808°F; 2100°C). The reason
for this choice is that the peak of the curve represents a radiation of 1.24 μm wavelength, which is equivalent to a frequency of 2.42 × 10^{14} cps. Planck's constant associated the quantum level of a photon of wave length 1.24 μm as having an energy level of 1 eV.

From both Figures 2.9 and 2.10, it appears that carbon disulfide can be autoignited at a low energy level of 0.15 to 0.18
eV (AIT = 194°F; 90°C) and that carbon monoxide can be autoignited at higher energy level of about 0.32–0.40 eV (AIT = 1128°F;
609°C). Thus, as the initial temperature is raised, we will
have crossed the various energy levels after which autoignition
commences for the various types of flammable chemicals. This
phenomenon is represented by the area under the curve and on
the right side of the peak. Because all of the molecules were in
a full state of agitation, autoignition became a reality. Combustion evidently begins as a relatively small reaction involving

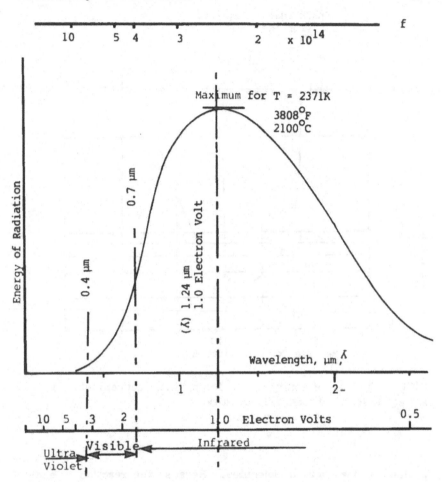

FIGURE 2.10 Energy spectrum for a specific temperature versus electron volts and radiative wave length. Frequency level (f) of electromagnetic radiation (cycles per sec; cps).

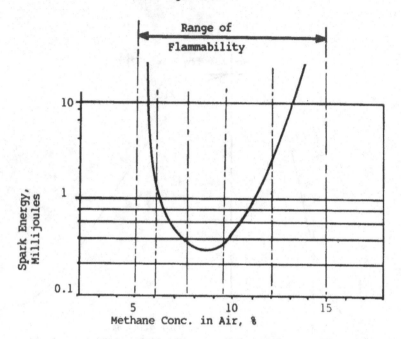

FIGURE 2.11 Ignitability curve for methane concentration in air at standard atmospheric conditions.

perhaps a dozen or so molecules. Because the reaction is exothermic, the pace of the reaction increases rapidly. This short time lapse explains why explosions do not develop instantaneously but, rather, have a short buildup time, a factor that is used in explosion detection devices. The radiation spectrum, shown in Figure 2.10, for a temperature of 2371°K (3808°F; 2908°C) therefore appears to truly represent the energy radiation as spread across a wide range of electron volts, with the bulk of the radiation being in the infrared range, a small amount being in the visible range, and only a very small amount being in the ultraviolet range. Figure 2.11 reveals a situation that is the cause of some of the ambiguities that exist in the conduct of ignition tests conducted by the US Bureau of Mines. The problems arise in the test apparatus itself. Usually, the combustion is carried out in 1-in. diameter Pyrex tubes, equipped with a spark igniter located in the base, with the flaming reaction occurring in an ascending manner. By this method the so-called ignitability limits are established that relate spark ignition energy to concentration of the fuel gas (here, methane) in air. The main interest is in that section of the curve below 10 mJ that spans methane concentrations between 6 and 13%. To ignite methane concentrations up to the full extent of the range of flammability, namely, 5 and ᷃15%, would require much more powerful ignition energies (over tenfold).

3
Flammability Limits of Mixtures of Two or More Flammable Gases

In the foregoing chapter, our use of the term "mixture" was confined to only one flammable gas that was mixed with air, to various extents, with resulting upper flammable (UFL) and lower flammable (LFL) limits. The air, itself, is a mixture comprising

    20.9% oxygen   (O_2)
    78.1% nitrogen   (N_2)
     0.9% argon   (A)
     0.1% miscellaneous, principally carbon dioxide (CO_2)

Frequently, however, we are also concerned with complex mixtures of two or more flammable gases

1. In pure condition mixed only with air.
2. Mixed with noncombustible gases and their effect on the flammability limits when mixed with air.

By applying methods that are used by the US Bureau of Mines, as outlined in their Bulletin 503 by Coward and Jones, in establishing the upper and lower limits of flammability, we can proceed as shown in the following examples.

The method that we will use follows the formula, set forth by the French physicist Le Chatelier, that was deduced from the observation that lower-limit individual mixtures, if mixed in any proportions result in mixtures that are also at their lower-limits. The same holds true for upper-limit mixtures.

COMBUSTIBLE GAS MIXTURES IN AIR ONLY

Example 1 Let us assume that it is required to establish the lower and upper limits of flammability of a "natural" petroleum gas mixture that has the following composition:

		%	LFL (%)	UFL (%)
CH_4	Methane	78	5.0	15.0
C_2H_6	Ethane	16	3.2	12.5
C_3H_8	Propane	4	2.2	9.5
C_4H_{10}	Butane	2	1.9	8.5

The general formula for determining the LFL of the mixture is

$$LFL = \frac{100}{P_1/LFL_1 + P_2/LFL_2 + P_3/LFL_3 + \cdots}$$

where P_1, P_2, P_3, and so on, represent the volumetric individual flammable gas percentages so that $P_1 + P_2 + P_3 + \ldots = 100$ (no air or inert gases present) and where LFL_1, LFL_2, LFL_3, and so on, represent the individual lower flammability limits; hence,

$$LFL \text{ (of mixture)} = \frac{100}{78/5 + 16/3.2 + 4/2.2 + 2/1.9}$$

$$= 4.3\% \text{ when mixed with air}$$

Similarly, the UFL of the mixture can be obtained by substituting the individual values of the upper flammability limits in the general formula; hence,

UFL (of mixture) = $\dfrac{100}{78/15 + 16/12.5 + 4/9.5 + 2/8.5}$

= 14.1% when mixed with air

For the examples that follow, we will, for the purposes of simplicity, concern ourselves with only three flammable gases and two inert gases, namely:

Hydrogen (H_2)
Carbon monoxide (CO)
Methane (CH_4)
Nitrogen (N_2)
Carbon dioxide (CO_2)

COMBUSTIBLE GAS MIXTURES IN AIR AND AN ADMIXTURE OF NONCOMBUSTIBLE GASES

Example 2 Determine the lower and upper flammability limits of a "producer" gas that has the following composition:

		%	LFL (%)	UFL (%)
CO	Carbon monoxide	27.3	12.5	74.0
H_2	Hydrogen	12.4	4.0	75.0
CH_4	Methane	0.7	5.0	15.0
N_2	Nitrogen	53.4	Inert	
CO_2	Carbon dioxide	6.2	Inert	

Now, because the mixture contains two inert gases, it will be necessary to dissect the mixtures into simpler mixtures, each of which contains only one flammable gas and part or all of the nitrogen or carbon dioxide, in such a manner to completely account for all of the inert gases. Any residual flammable gas fraction that is not paired with any remaining inert gas will be treated as shown in the previous example. Refer now to Figure 3.1 and apportion the inert gas and associate it with a convenient flammable gas as described previously. The dissection, al-

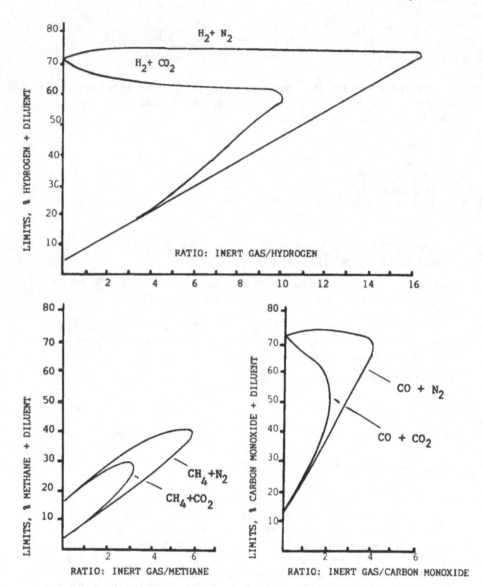

FIGURE 3.1 Flammability limits of hydrogen, carbon monoxide, and methane with various amounts of carbon dioxide and nitrogen.

though it may appear arbitrary, will always result in the same end values. Therefore, let us pair the gases as follows:

$$27.3\% \text{ CO} + 6.2\% \text{ CO}_2 = 33.5\% \tag{1}$$

$$12.4\% \text{ H}_2 + 53.4\% \text{ N}_2 = 65.8\% \tag{2}$$

$$0.7\% \text{ CH}_4 \qquad\qquad = 0.7\% \tag{3}$$

Note, from Figure 3.1, that there are three sets of curves for an individual flammable gas mixed with either of two inert gases. First, establish the ratio of the inert gas to the combustible gas with which it is associated. Then, by referring the appropriate curve in Figure 3.1, establish the flammable limits for each of the preceding equations, as follows:

Equation	Inert/combustible gas ratio	Lower limit	Upper limit
(1)	0.2	17	70
(2)	4.3	23	76
(3)	0.0	5	15

By combining the individual limits and using the generalized form of the Le Chatelier formula:

$$\text{Lower flammable limit} = \frac{100}{\left(\dfrac{33.5}{17}\right) + \left(\dfrac{65.8}{23}\right) + \left(\dfrac{0.7}{5}\right)}$$

$$= 20.2\%$$

$$\text{Upper flammable limit} = \frac{100}{\left(\dfrac{33.5}{70}\right) + \left(\dfrac{65.8}{76}\right) + \left(\dfrac{0.7}{15}\right)}$$

$$= 73.5\%$$

Example 3 Determine the upper and lower flammable limits for gases obtained from a coal mine after a fire or explosion. The usual practice in such circumstances is to seal off the mine entrances and not to reenter until conditions become safe. Samples of the mine atmosphere were taken, and one such sample had the following composition:

		%	Air-free basis (%)	LFL (%)	UFL (%)
CO	Carbon monoxide	4.3	5.0	12.5	74.0
H_2	Hydrogen	4.9	5.7	4.0	75.0
CH_4	Methane	3.3	3.8	5.0	15.0
N_2	Nitrogen	70.9	69.6	Inert	
CO_2	Carbon dioxide	13.8	15.9	Inert	
O_2	Oxygen	2.8	0.0	Gas sample was not air-free	

Because the gas sample contained a small amount of oxygen, a
result of a small amount of air being present, the first step is
to recalculate the air sample on an air-free basis. The amount
of air present in the original sample is 2.8 × (100/20.9) = 13.4%.
The recalculated values are tabulated upon the basis of the air-
free sample being 86.6% of the original volume. The percentage
of nitrogen is the difference between 100 and the sum of the
other gases.

The procedure now closely follows that employed in Example
2; therefore, let us pair the gases as follows:

$$5.0\% \text{ CO} + 17.5\% \text{ N}_2 = 22.5\% \tag{1}$$

$$3.8\% \text{ CH}_4 + 20.9\% \text{ N}_2 = 24.7\% \tag{2}$$

$$3.0\% \text{ H}_2 + 31.2\% \text{ N}_2 = 34.2\% \tag{3}$$

$$2.7\% \text{ H}_2 + 15.9\% \text{ CO}_2 = 18.6\% \tag{4}$$

It will be noted that the hydrogen constituent had to be paired,
partly with the residual nitrogen that remained after having
been paired with the carbon monoxide and the methane. Hence,
the remaining hydrogen was required to be paired with the carbon dioxide. Thus, all of the inert gas constituents were accounted for.

Equation	Inert/combustible gas ratio	LFL (%)	UFL (%)
(1)	3.5	61	73
(2)	5.5	36	42
(3)	10.4	50	76
(4)	5.9	32	64

$$\text{Lower limit (air-free mixture)} = \frac{100}{22.5/_{61} + 24.7/_{36} + 34.2/_{50} + 18.6/_{32}}$$

$$= 43\%$$

$$\text{Upper limit (air-free mixture)} = \frac{100}{22.5/_{73} + 24.7/_{42} + 34.2/_{76} + 18.6/_{64}}$$

$$= 61\%$$

Because the air-free mixture is 86.6% of the whole original mixture, the limits of the mine fire atmosphere are

$$\text{Lower limit (LFL)} = 43 \times \left(\frac{100}{86.6\%}\right) = 50\%$$

$$\text{Upper limit (UFL)} = 61 \times \left(\frac{100}{86.6\%}\right) = 70\%$$

(*Source*: U.S. Bureau of Mines Bulletin 503 by Coward and Jones, pp. 5–8, for Example 3.)

RENDERING AN ENCLOSURE NONFLAMMABLE IN THE PRESENCE OF A FLAMMABLE GAS AND AIR BY THE USE OF AN INERT GAS

To prevent fires or explosions from occurring in enclosed spaces, fuel tanks, chemical processing equipment, pipelines, and such, it is most obvious that either air purging, or the injection of an inert gas followed by air purging will be required if personnel are to enter the space. The process of inactivation is best described by referring to Figure 3.2. Herein, the flammable gas content is plotted versus the volume percent of

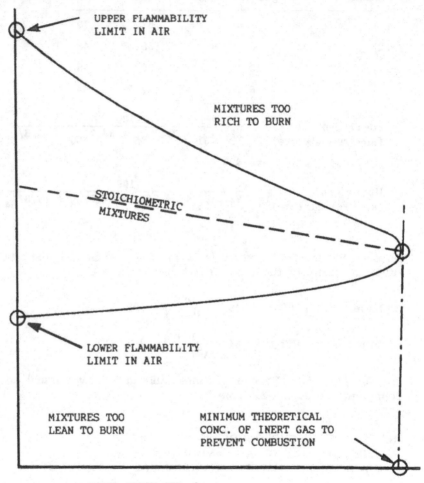

FIGURE 3.2 Typical generalized flammability limits of a combustible gas diluted with various amounts of an inert gas.

the added inert gas. Note that when no inert gas has been added the "shoe-shaped" curve intersects the left-hand ordinate at the lower flammability limit and at the upper flammability limit in air. As the percentage of inert gas is increased, the lower and upper flammability portions of the curve converge, raising the lower and depressing the upper flammability limits. Where the curves converge is at the theoretical minimum concentration of inert gas needed to prevent combustion. The only combustion that can occur is within the confines of the flammability curves. As previously defined, the stoichiometric mixture lines represent those proportions of flammable gas and air that are theoretically needed for complete combustion, with no excess fuel or air remaining after the reaction. With most fuels, the LFL curve rises much less sharply than the descending UFL curve. Figure 3.2 is shown for generalized purposes and does not represent any particular flammable gas or vapor.

Getting down to fundamentals, let us examine the effects produced when methane gas is mixed, to various extents with four very different gases, to inhibit the formulation of flammable mixtures (Fig. 3.3). Every flammable gas or vapor has its own specific and peculiar properties. There is no way to predict how some specific gas will behave. We must rely on experimental data that have been obtained mostly by the US Bureau of Mines and are outlined in Bulletins 503 and 627, which constitute the fountainhead of information. Figure 3.3 illustrates the wide differences that result when establishing the minimum theoretical concentration of each of the four inert gases to prevent combustion of methane, namely:

1. Halon 1301 (bromotrifluoromethane; CF_3Br). The extraordinary low percentage required (2%) is due to this compound's characteristic flame-extinguishing ability, which will be discussed in detail in Chapter 5.
2. Carbon dioxide is the usual gas that is used, assuming that is does not react unfavorably with the contents of the space. A 25% concentration is required.
3. Water vapor in the form of steam can be used at a minimum concentration of 29%. It is a requirement that the space is sufficiently hot (over 160°F; 70°C) to preclude condensation. This should not present a problem where steam, even at low pressures, is available.
4. Nitrogen gas has been used when, for some reason, carbon dioxide is either not available or not compatible with

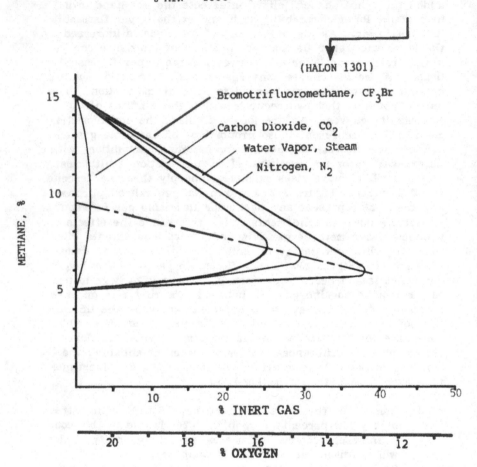

FIGURE 3.3 Flammability limits of methane in separate mixtures of inert gas and air.

the contents of the space. The required nitrogen concentration is 38%.

Note, in all of the preceding examples, we are referring to standard atmospheric conditions as illustrated and emphasized in Figure 2.2 (Chap. 2), in which adjustments to flammable limits are necessary at temperatures other than standard.

With the exception of the first inert gas (CF_3Br) used for inactivation, the other three gases substantially reduce the oxygen content but to noticeably different degrees. The reasons for this disparity have been ascribed to three very probable generalized causes: the molar specific heat; thermal conductivity; and, the third, although somewhat vague, is probably the most critical, it involves the temperature developed at the source of ignition. The heat of combustion of a limiting mixture of fuel vapor and air is enough to raise the surrounding unburned mixture to a temperature far in excess of its ignition temperature, as usually determined, but probably it is not high enough to permit inflammation to proceed, considering the borderline mixture conditions within the short duration of attempted ignition. One can conclude, perhaps, that "free radicals" (unstable molecular fragments with high chemical activity) are formed increasingly as the boundary temperature of the incipient flame is increased. The free radicals of fuel gas, or vapor, variously interact with the different inert inactivating agents. It is not well understood but, for a certainty, it involves more than the reduction of the oxygen content above the area. Table 3.1 is a condensed summary of flammability limits for six common pure fuel gases and vapors plus an average gasoline, which is a mixture of a number of separate vapors. In reviewing the array of values, one sees a number of anomalies, particularly for the percentage of oxygen present under different conditions for forming an inert atmosphere. In effect, we are admitting our inability to attribute, with certainty, any cause for the anomalies until more research is conducted. The ratios of the percentage of inert gas to the percentage of flammable gas would, at first glance, appear to be a promising index. However, the computed percentage of carbon dioxide appearing in the last column does agree with the carbon dioxide concentrations shown in Table 3.2, which is used as a guide for the design of carbon dioxide systems. This should be no surprise, because the data were derived from the US Bureau of Mines

TABLE 3.1 Some Typical Flammability Limits and Corresponding Minimum Inert Gas Requirements

| | Flammability limits in air (no inert gas) | | | | Minimum theoretical inert gas mixture to prevent combustion | | | | |
| | LFL | | UFL | | Nitrogen (N_2) | | Carbon dioxide (CO_2) | | |
Gases	%	O_2 (%)	%	O_2 (%)	N_2/flam. gas ratio	%	CO_2/flam. gas ratio	O_2 (%)	CO_2 (%)
Hydrogen (H_2)	4.0	20.8	75.0	5.2	16.5	5.2	10.2	8.3	61
Carbon monoxide (CO)	12.5	18.3	74.0	5.3	4.0	6.0	2.0	9.1	52
Methane (CH_4)	5.0	19.8	15.0	17.8	6.0	12.5	3.3	15.0	25
Ethane (C_2H_6)	3.0	20.3	12.4	18.3	12.8	11.2	7.2	13.5	33
Ethylene (C_2H_4)	2.7	20.5	30.6	10.0	15.6	10.0	9.0	12.3	41
Benzene (C_6H_6)	1.4	20.6	7.1	11.7	21.2	11.7	13.0	14.6	31
Average gasoline	1.4	20.6	7.1	11.7	a		a		29

aA varying mixture (see Fig. 3.4).

TABLE 3.2 Minimum Carbon Dioxide Concentration for Extinguishment or Inactivation

Compound	Theoretical	Design
Hydrogen	62%	74%
Acetylene	55	66
Carbon disulfide	55	66
Carbon monoxide	53	64
Ethylene oxide	44	53
Ethylene	41	49
Ethyl ether	38	46
Ethyl alcohol	36	43
Butadiene	34	41
Ethane	33	40
Benzol	31	37
Cyclopropane	31	37
Propane	30	36
Propylene	30	36
Isobutane	30	36
Pentane	29	35
Hexane	29	35
Gasoline	28	34
Kerosene	28	34
Quench, lube oils	28	34
Butane	28	34
Methyl alcohol	26	31
Acetone	26	31
Methane	25	30
Ethyl dichloride	21	25

Bulletin 503 that was previously mentioned. The guide for the
design of carbon dioxide extinguishing systems, namely the National Fire Codes, NFPA Standard 12, has been in effect for
over 50 years and has been checked and confirmed many times.
Table 3.2 has been the standard wherein the theoretical minimum concentration of carbon dioxide has been increased, by 20%
over the minimum calculated concentration, as the actual value
to be used that includes a safety factor. The 25 different materials are listed in descending order by the degree of carbon
dioxide concentration required and are representative of a wide
range of flammable chemicals. They are pure materials, with
the exception of gasoline, kerosine, quench and lubricating oils,
which are mixtures that would embrace naphtha, mineral spirits,
diesel fuel, aviation jet fuel, fuel oils, transformer oils, and
the like, as well. Figure 3.4 illustrates the span of various
petroleum products in terms of the number of carbon atoms in
the molecules. We must remember that each category for the
number of carbon atoms includes both normal and isomeric con-
figurations. With gasoline, for instance, gas chromatograph
charts show from 16 to 18 different materials. Commercial liquid fuels are invariably blends of many related materials, with
either the same or closely related inactivation requirements.

Figure 3.5 shows the mathematical relationship, in a generalized form, for determining the percentage of any inert gas existing at any time (t), within a space volume (V), when being
injected at a rate (a), while air is flowing through the space at
a continuous rate (b). To solve this problem, it is necessary
to make the assumption that gaseous diffusion occurs immediately and that the percentage of the inert gas is the same at all
points within the enclosure at any given moment. The inert
gas must always be applied in a manner that promotes progressive mixing of the atmosphere within the enclosed space. The
displaced atmosphere is freely exhausted through various small
openings, or through special vents, as the inert gas is injected.
Inevitably therefore, some of the inert gas is lost, which increases at high concentrations. This method of application is
called "free-efflux" flooding.

It is obvious that to rapidly reach the desired inert gas concentration and to retain that percentage as long as practically
possible would require that the airflow (b) be eliminated. The
equation now becomes considerably simplified as shown on the
bottom of Figure 3.5. These conditions are what a conventional

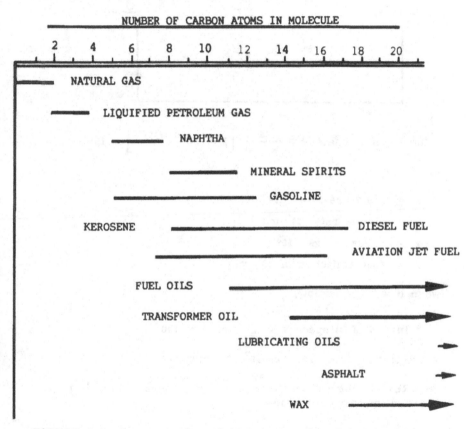

FIGURE 3.4 Summary of product types produced from petroleum.

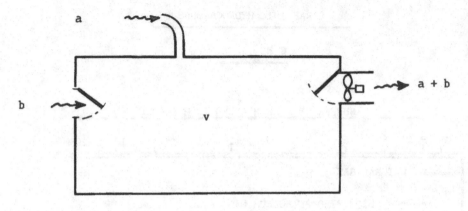

$$\% \text{ Inert gas in space } V = \frac{a}{a + b} \left[1 - e^{-(a+b)t/V} \right] \times 100$$

where:

     a = inert gas flow rate, ft^3/min

     b = airflow rate, ft^3/min

     v = space volume, ft^3

     t = time (injection of a), min

when $b = 0$ (no airflow)

$$\% \text{ Inert gas in space } V = (\dot{i} - e^{-x}) \times 100$$

where $x = ft^3$ inert gas per ft^3 of volume (V)

FIGURE 3.5 General formula: inert gas concentration in a confined space versus time.

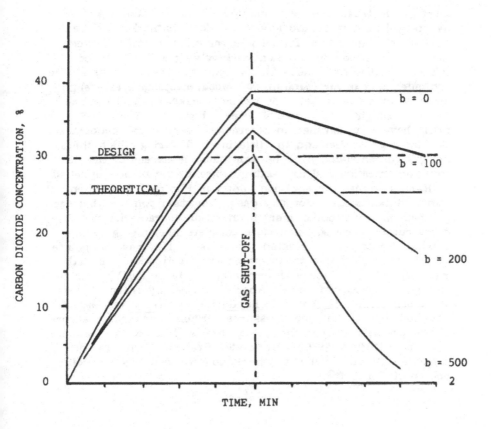

EXAMPLE:

Flammable gas hazard - Methane

Space volume (V) - 1000 ft^3

Inert gas - Carbon dioxide

Inert gas, flow rate 500 CFM (a)

Inert gas flow duration - 1 min (t)

AIRFLOW OF AIR (0–500 CFM) (b)
   (CONTINUOUS)

FIGURE 3.6 Effect of various airflows on inert gas concentration for the example shown.

inert gas installation usually includes. All air intakes and ex-
hausts will be closed, and any mechanical ventilation will be
shut down. As is true for all general rules, there are excep-
tions. I remember one application involving a "hot" cell containing a radioactive fuel. Safety requirements necessitated the
continued flow of air through the space, keeping it at a slight
negative pressure to preclude external leakage. The air was
passed through a very special filter. Figures 3.5 and 3.6
would have to be studied to determine the inert gas concentration for a given fuel and the time interval during which the inert gas concentration is to be maintained in excess of the design concentration and fire extinguishment must be accomplished.

Remember that the equation shown in Figure 3.5 is a general
form and can be used for any gas. The main point is that each
gas has its own peculiar inert inactivation characteristics. Because our data are so much more complete when using carbon
dioxide, we have concentrated on its use. This, as was pointed
out, is principally due to its high inert inactivation and extinguishing ability. Bromotrifluoromethane (Halon 1301) is unusual
and will be discussed later. Another compelling reason for use
of carbon dioxide is that it is handled in commerce in compact
liquefied form at elevated pressures. When expanded to atmospheric pressure through horns or nozzles, it reaches a temperature of $-110°F$ ($-79°C$) and, under fire conditions, rapidly
expands. At standard atmospheric conditions it has a specific
volume of 8.6 lb/ft^3.

4
Pyrolytic Decomposition
of Fuel Vapors and Combustion
Chain Reactions

Organic chemistry is the chemistry of the compounds of carbon.
Even after it had become evident that these compounds did not
have to originate in living processes, as once thought, but could
be synthesized, it was convenient to retain the name "organic"
to describe them and related compounds. Coal and petroleum
provide two large reservoirs of organic material from which the
simpler compounds can be derived that, when used as building
blocks, will provide even larger and more complicated compounds.
Each different assembly of atoms has its own characteristic set
of chemical and physical properties. Well over a million of these
compounds exist and the list is constantly expanding.

What is so special about the element carbon is its ability to
attach to other carbon atoms to an extent that is both unique
and not possible for any other element in nature. Their aggre-
gation can form chains of all lengths and rings of all sizes,
both of which can have branches and cross-links. Other atoms,
chiefly hydrogen, are attached to these chains and links, but
the halogens, oxygen, nitrogen, sulfur, and many more, may
also be attached. In this text we can concern ourselves to only one limited, but very important, part of this enormous subject.

A system of nomenclature has been devised for all of the known compounds, which is known as the International Union of Pure and Applied Chemistry (IUPAC) system. In this text the various organic chemical compounds are so identified.

The organized study of chemistry began about 2 centuries ago. The first objects of consideration concerned combustion because of its immense importance to man and his survival. The mechanism of the combustion reaction is extremely complicated and, although great strides have been taken, the reaction is not, as yet, fully understood. We always try to explain "why" organic compounds react as they do and are not satisfied with a qualitative description of "how" they react. Although we wish to be as exact as possible, our lack of a fuller understanding forces us to compromise. However, of one thing we are certain, namely, that flaming combustion is a free-radical chain reaction. The combustion reaction, from an overall viewpoint, is extremely exothermic, and yet for almost all organic fuels it requires (1) a high, localized temperature, such as a flame or a spark, for its combustion and propagation, or (2) that a sufficiently high uniform temperature distribution (autoignition) exists, per se, to initiate combustion.

An endothermic requirement for the combustion process is the preparation necessary to condition the fuel vapors to participate in the reaction. This requirement is over and beyond that required, as shown in the frontispiece, to vaporize the liquid fuel or to pyrolytically decompose the solid fuel to rise into the fire area. Fuel gases and vapors that are in their natural state require no such preparation. The endothermic requirement for the combustion process involves the following:

1. Fuel molecules do not just burn "in toto".
2. The fuel molecules must be "disentangled" under elevated temperature conditions, wherein the weakest interatomic bonds become dissociated first.
3. Relatively large molecules are fractured into smaller molecules and molecular fragments (free radicals) having high chemical activity, such as (CH_3^*), (CH_2^*), H^*, and C.
4. The free elemental carbon that is released burns slowly compared with the other free radicals and, hence, is the last element to burn.
5. Because of this apparent slowness, the carbon will appear as black smoke above the zones of combustion.
6. The opacity of the smoke is a function of the percentage of carbon within the fuel molecule.

PYROLYTIC DECOMPOSITION

The term, pyrolytic decomposition, defines the decomposition by the action of heat alone. The word pyrolysis is taken from the Greek *pyr*, fire plus *lysis*, a loosing and, hence, to chemists means "cleavage by heat": compare the word with hydrolysis, as "cleavage by water."

Much of what we know about pyrolytic decomposition has been obtained from petrochemical research. The pyrolysis of the *aliphatic* type of hydrocarbons, including both "open-chain" and "ring" molecular structures is known in the petroleum industry as "cracking" and is the source of a huge number of various types of organic chemicals. In these industrial applications, cracking is carried on only as far as is required to obtain certain end products.

Cracking is the process of converting large molecules into smaller ones by the application of heat, either alone or with the aid of catalysts. The older process, using heat alone, consists of passing the material to be treated through a chamber heated to between 900 and 1100°F, with pressures from 600 to 1000 psi. With the use of catalysts, consisting of various metallic oxides, and at about the same temperatures but at pressures only slightly higher than atmospheric, an equivalent yield is possible. These processes, and particularly the latter, have been acclaimed as the greatest force for conservation that was ever developed by the chemical industries.

The products of the cracking process are the smaller structure alkanes (C_nH_{2n+2}) and the simpler alkenes (C_nH_{2n}), plus some hydrogen. Thus, the cracking of the mixture of compounds in the crude petroleum leads to a multitude of possible products such that the representation of the process by simple equations is impossible. The heat of decomposition is highly variable; if large amounts of gas are produced, the heat of decomposition is highly endothermic and relatively high. Conversely, if the heat input is kept relatively low, the gas production would be virtually eliminated. Further study of other organic chemistry texts is recommended for readers who wish to pursue this subject in some depth. This book can serve only as an overview of this most complex subject.

Figure 4.1 represents a classification system for hydrocarbon gases and liquids. This grouping, although limited, represents what can be considered most of those flammables that are both widely used and possess the greatest heat content in fire situations. Note, each of the five vertical groupings are identified as alkanes, alkenes, alkynes, cyclics, and arenes, and each

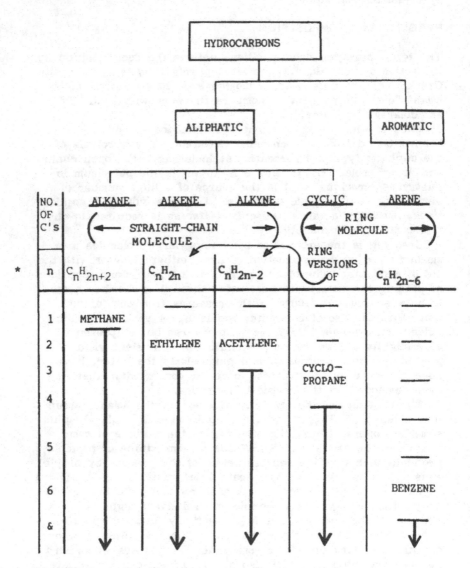

FIGURE 4.1 Classification system for hydrocarbon gases and liquids.

forms a homologous series. A further additional identification
is provided by attaching a prefix indicating the number of carbon atoms in the molecule: hence, a C_5 alkane, more familiarly,
is named pentane, or a C_7 arene, more familiarly, is named tol-
uene. The listings are headed by only the first and simplest
compound in each category and are further identified by a generalized formula for each homologous series. With the exception
of the liquid, benzene, in the arene series, each of the other
four categories are headed by a gas. Usually, when the carbon count becomes 5 or more the compound becomes a volatile
liquid (under normal atmospheric conditions). As the carbon
count increases, the liquids gradually become oily and increasingly viscous, and around C_{20} they become waxy solids. Observe, that the cyclic aliphatics are ring versions of the alkene
and alkyne straight-chain molecules and have the same numbers
of carbon and hydrogen atoms for a given horizontal grouping.
Note, particularly, that the difference, molecularly speaking,
between any adjacent horizontal grouping *and* also within any of
the five homologous series is equal to CH_2, the methylene free
radical. This will become apparent by examining the generalized formula for each homologous series, for which the addition
(or subtraction) of one carbon atom is accompanied by the addition (or subtraction) of two hydrogen atoms. Hence, if pentane
(alkane homologous series) is cracked, the result is for the
pentane to yield a CH_2 (methylene) radical that, incidentally,
is a building block for the formation for the alkene series,
which are polymerized groups of methylene radicals. However,
large molecules are more susceptible to being fractured by high
temperatures and, thus, as the alkane molecule is shortened,
the alkene molecule is lengthened. This process will terminate
when the original molecules are completely fractured into a mixture of highly active free radicals including, CH_3^*, CH_2^*, H^*,
C, and H_2. These are now the gaseous fuels that the fire di-
gests, and they exist in a rapidly ascending force within the
cracking zone shown in Figure 1.3. Air is not present in this
latter zone because it was consumed in the flaming envelope
that envelops the cracking zone.

The foregoing example of the cracking process is perhaps
oversimplified and idealized; but we can be certain that for
flaming combustion, as is generally portrayed in Figure 1.3, *all*
fuels, to burn, must be volatilized *and* cracked for the combustion process to be initiated and made continuous. This is true

no matter what the original fuel may have been. The frontispiece will show that even for the burning of natural gases, a
cracking interlude is necessary for combustion to continue.

COMBUSTION CHAIN REACTIONS

It is important for us to know not only what happens in a chemical reaction but also how it happens, that is, to know not only
the facts but also the theory. A detailed step-by-step description of a chemical reaction is called a "mechanism," yet it is
really only a hypothesis and is advanced to account for the
known facts. It would be difficult to say that a mechanism had
ever been proved. If, however, a mechanism accounts satisfactorily for a wide variety of facts; if we make predictions based
upon this mechanism and find them borne out; if the mechanism
is consistent with mechanism for other related reactions; then
the mechanism is said to be established, and it becomes a part
of theoretical chemistry. An understanding of these mechanisms
will assist us to see a pattern in the confusing multitude, particularly, of organic reactions. Our interest will be confined to
combustion chain reactions in particular that would promote oxidation as well as the reverse effect of introducing inhibitive
chain reactions to accomplish fire extinguishment, which will be
discussed in detail in Chapter 6 under fire extinguishing mechanisms. As illustrations of actual examples, four exothermic oxidation reactions are represented by diagrams (Figs. 4.2, 4.3,
4.4, and 4.5).

The Basic Equation $Cl_2 + H_2 = 2HCl$

The reaction of chlorine with hydrogen to form hydrochloric
acid is strongly exothermic and has been previously covered in
Chapter 2, but it will be repeated here from the standpoint of
progressive chain reactions. Although this reaction does not
involve oxygen, as such, it nevertheless is an oxidation-type
reaction and is completely analagous to combustion.

Among the facts that must be accounted for are that (1) hydrogen and chlorine when combined do not react in the dark at
room temperature; (2) the reaction takes place readily, however,
in the dark when the temperature exceeds 500°F (260°C); (3)
under the influence of the ultraviolet portion of sunlight, the
reaction takes place readily in an explosive manner; (4) the ve-

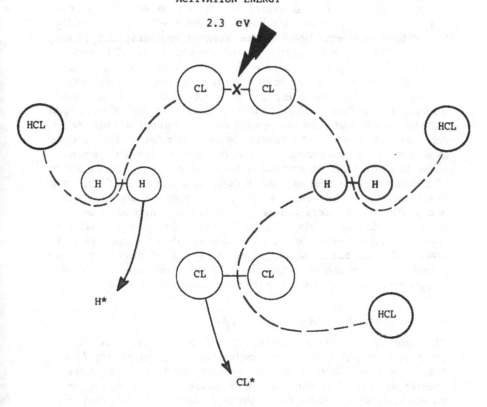

BASIC EQUATION $CL_2 + H_2 = 2HCL$

ACTIVATION ENERGY

2.3 eV

$$CL_2 + e = CL^* + CL^*$$

$$CL^* + H_2 = HCL + H^*$$

$$H^* + CL_2 = HCL + CL^*$$

ETC.

FIGURE 4.2 Example of an unbranched-chain reaction (chlorination of hydrogen).

locity of the reaction is influenced by the degree of sunlight to which it is exposed and is considerably subdued in a proportionate manner. Refer to Fig. 4.2, which is a prime example of a simple unbranched chain reaction. The first step is to break the weakest of the molecular bonds of the reactants (Cl_2 or H_2). The chlorine-chlorine bond is the weakest requiring 2.3 eVolts, versus 4.1 eVs for the hydrogen-hydrogen bond. The two chlorine free radicals possess a great deal of energy and have the ability to directly fracture a hydrogen molecule (H_2) into two hydrogen free radicals, one of which combines with the colliding chlorine free radical to form a final molecule of hydrogen chloride (HCl) *but* with the evolution of a hydrogen free radical. As the bottommost equation shows, this latter free radical is absorbed by the remaining chlorine free radical to form another final molecule of hydrogen chloride *but* with the evolution of a chlorine free radical, and so on, in a series of chain-propagating steps. The reaction being exothermic is self-sustaining and sufficient in released heat energy to continue more and more chain-initiating steps in which new chlorine molecules are fractured into free radicals, until the supply of molecules to be fractured runs out or some countermeasures (if any) can be employed. This example is frequently used to exemplify the unbranched-chain reaction because it is relatively simple.

The Basic Equation $2H_2 + O_2 = 2H_2O$

The reaction of hydrogen with oxygen to form water is strongly exothermic and is the fastest combustion reaction known (Fig. 4.3). By comparison with the previous example, it requires a greater amount of energy (4.1 eV) to fracture the hydrogen molecule, but this amount of energy is still less than that required to fracture an oxygen molecule (4.9 eV). Just as seen with the previous example, all that is required to start the reaction is to fracture a single molecule to trigger the avalanche. The exothermic nature of the reaction will sustain the process. This is an example of a branched-chain reaction. A close examination comparing this type of chain reaction with the unbranched-chain reaction shows that there are many more free radicals created.

After the fracture of a hydrogen molecule into two hydrogen free radicals, one of the latter, because of its high energy, fractures an oxygen molecule and takes an oxygen atom unto itself forming a hydroxyl free radical (OH^*) as well as an unat-

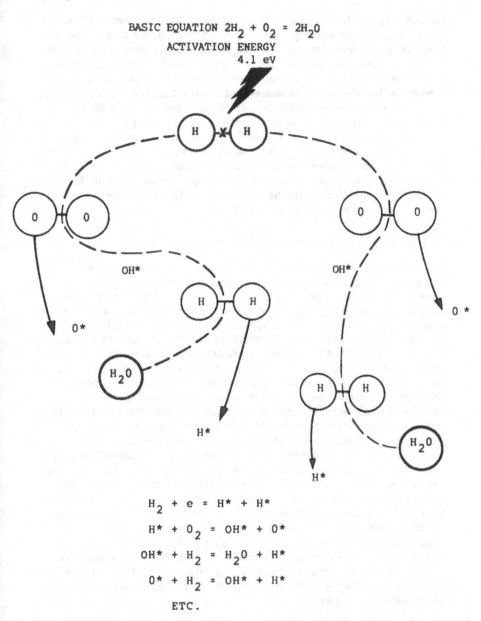

$$H_2 + e = H^* + H^*$$

$$H^* + O_2 = OH^* + O^*$$

$$OH^* + H_2 = H_2O + H^*$$

$$O^* + H_2 = OH^* + H^*$$

ETC.

FIGURE 4.3 Example of a branched-chain reaction (oxygen—hydrogen reaction).

tached oxygen free radical (O*). The hydroxyl free radical, which has a high energy, fractures another hydrogen molecule, thereby forming moisture and releasing a hydrogen free radical (H*) in exchange, and so forth. These chain-propagating reactions after the chain-initiating reaction will continue until they terminate for the same reasons that existed in the first example. Note, there are three different types of free radicals in the present reaction, all of which are extremely energetic.

Up to this point, it has been possible to illustrate the patterns of the reactions in a simple diagram. However, as we begin to become involved with carbon-containing fuels, the complexities increase and become increasingly speculative in illustrating the reaction patterns. However, there is strong evidence that the reasoning is sound.

The Basic Equation $2CO + O_2 = 2CO_2$

This example was chosen for two reasons: first, because it is commonly associated with all combustion phenomena and, second, because of its ignition requirements. It has frequently been stated that the reaction as depicted is nonexistent, except when moisture is present. The upper and lower flammability limits for carbon monoxide were determined by the US Bureau of Mines (refer to Bulletin 503) under conditions of 100% relative humidity and 70°F (21°C). When water vapor is removed as completely as possible in laboratory experiments, the most explosive (stoichiometric) mixtures of carbon monoxide and pure oxygen can be ignited only by unusually powerful electric sparks. The moisture content of 100% relative humidity amounts to 2% by volume, which is equivalent to a vapor pressure of 15.2 mmHg at 70°F (21°C). Even when the temperature of condensation (dew point) of the airborne moisture is reduced to 0°F (−18°C), the vapor pressure is lowered to 1.0 mmHg, the moisture content is still appreciable and amounts to about 0.1%. The point is that moisture is present all about us, in sufficient amounts to be of significance. By referring to Figure 4.4, you will note that Equation (1) starts the reaction by the dissociation of a moisture molecule and not by the dissociation of a carbon monoxide molecule. This is because the atmosphere that we are discussing contains basically the following molecules, with their dissociation energies:

BASIC EQUATION $2CO + O_2 = 2CO_2$

Activation Energy
4.1 eV

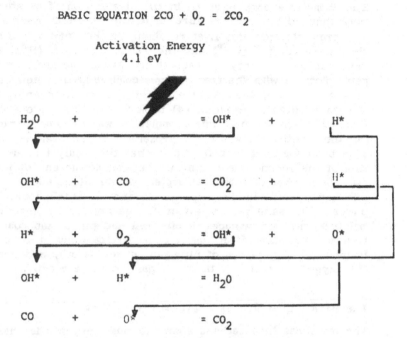

1   H_2O +   = OH*   +   H*

2   OH*   +   CO   = CO_2 +   H*

3   H*   + O_2 = OH*   +   O*

4   OH*   +   H*   = H_2O

5   CO   +   O*   = CO_2

FIGURE 4.4 Example of the oxidation of carbon monoxide.
Note that the presence of moisture is required for the basic
reaction to reach completion.

Carbon monoxide (CO)	7.8 eV
Atmospheric nitrogen (N_2)	6.3 eV
Atmospheric oxygen (O_2)	4.9 eV
Atmospheric moisture (H—OH)	4.1 eV

The dissociation energy of the carbon monoxide molecule is the
greatest of any diatomic molecule, and it is appreciably more
than that of the diatomic molecule N_2, in which two nitrogen atoms are joined by three covalent bonds. The lowest dissociation
bond energy then prevails in initiating the overall reactions.

The chemical energy released in the dissociation of moisture is more than adequate to fracture all of the molecular reactants. The products from this first reaction are the very active species, the hydroxyl (OH^*) and the atomic hydrogen (H^*), that have sufficient energy to fracture the carbon monoxide and oxygen molecules with the free radicals combining in a new order, as shown. Note, the moisture, which played the opening role, is reconstituted in Equation (4) back to moisture. Note also, that, although the presence of moisture was evident in motivating the reaction, it was not represented in the basic equation. This behavior is truly catalytic in that the catalyst altered the rate of the reaction, emerging unchanged at the end of the process. A number of similar examples can be cited, such as water injection into the intake of piston engines to obtain increased power. The same general effect can be noticed by injecting moisture into the furnaces of oil-fired and pulverized coal-fired boilers. Moisture affects many chemical reactions in a similar fashion. The presence of nitrogen serves as only a diluent for the oxygen content. It takes no part in these reactions.

The Basic Equation $CH_4 + 2O_2 = CO_2 + 2H_2O$

The inertness that methane shows to most reagents is characteristic of the alkane structure (chainlike) in general. Methane is the most heat-stable of all the hydrocarbons, having an AIT of about $1000°F$ ($788°C$). As the homologous series progresses toward larger molecules, the temperature limits become lower. Like methane, the higher alkanes undergo comparatively few reactions but, when they do take place and then only under vigorous conditions, they yield mixtures of products. Combustion reactions of the free radical type are typical of what can be termed vigorous.

Figure 4.5 illustrates both the oxidation steps involved in the combustion of methane and the burning pattern of all hydrocarbons, which after cracking, are largely reduced to methane and similar compounds. The oxidation of methane takes a relatively large number of steps. Equation (3) is particularly recognizable as the formation of methyl alcohol, the first stage of the methane oxidation. Equation (4) is recognized as the formation of formaldehyde. Equation (6) can be recognized as representing the formation of formic acid, and the terminal Equa-

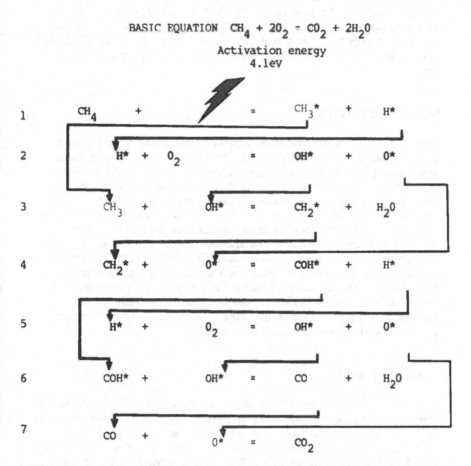

FIGURE 4.5 Example of the oxidation of methane.

tion (7) is completed by formation of the last oxygen free radical. In the event of incomplete combustion, the sequence becomes interrupted, and the presence of formaldehyde and formic acid is noticeable because of their pungency. They are particularly noticeable in fire residues by their characteristic odor. The common denominator for all of these chain reactions involving the concomitant presence of H^*, O^*, and OH^* free radicals has caused this interaction to be named the "hydroxylation theory."

FLAME VELOCITIES

Figure 4.6 shows the relationship between flame velocity and the gas—air mixture for each of the three oxidation reactions described previously (see Figs. 4.3—4.5). It should come as no surprise that those reactions that have the largest "population" of free radicals are faster than those with fewer: hydrogen has the greatest flame velocity and methane the least flame velocity. The presence of incandescent carbon atoms observable in the flames, reduces the flaming velocity of methane and its related homologues. Flame velocities are matters of importance to the designers of gas-fired apparatus, particularly, as they pertain to the ease of flame "low out." The numerical values of flame velocity are all based upon the Bureau of Mines testing procedure using 1-in. diameter tubes. Actual unconfined velocities are several times larger.

FLAME FLICKER

A most valuable observation, made by photometric equipment of the luminous and nonluminous radiation from the flames of burning hydrocarbons, reveals the phenomenon of flame "flicker." The flicker occurs between 10 and 40 cps, with the maximum effect being noticed at 25 cps. This phenomenon is put to use by manufacturers of infrared fire detection equipment as a means of discriminating flames from other sources of infrared emanation such as sunlight, incandescent lamps, and the like.

The upper section of Figure 4.7, relates the intensity of flicker to the observed frequency in cycles per second. Flicker further indicates the presence of a vibrational phenomenon

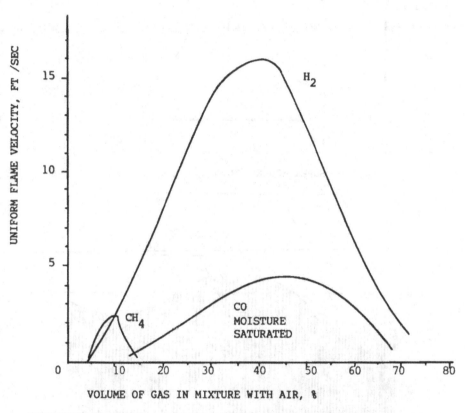

FIGURE 4.6 Uniform flame velocity of hydrogen, carbon monoxide, and methane versus air/mixture ratio. Basis: upward velocity in a 2-in. tube at standard atmospheric conditions (from US Bureau of Mines Technical Paper 427 and International Critical Tables).

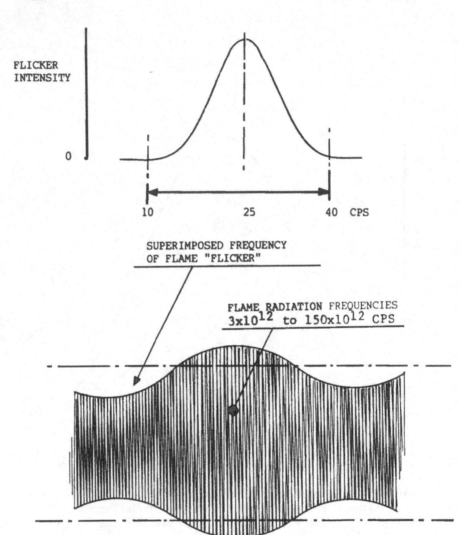

FIGURE 4.7 Flame radiation modulation (flicker) as observed from combustion of hydrocarbon gases and vapors.

wherein energy is constantly and periodically being transferred from one extreme state to another. Whether or not the vibration is harmonic is unimportant. The lower portion of Figure 4.7 shows how these flicker frequencies can be superimposed over the basic whole-flame radiation frequency that ranges from 3×10^{12} to 150×10^{12} cps and varies from the ultraviolet, through the visible range, into the long-infrared region. This phenomenon is analogous to a radiocommunication that is modulated by superimposed high-frequency signals. It is no doubt speculative to attribute this phenomenon to the interplay of the various free radicals during the chain-reaction process and to the presence of incandescent carbon atoms; however, no other explanation seems possible. The flicker behavior is now known to exist for hydrocarbon fuels. More research is necessary to investigate the other organic fuels such as alcohols, ketones, aldehydes, ethers, esters, and so on. Furthermore, it is interesting that this phenomenon has been known to exist for over a century, because the flicker frequency lies within the lower range of human auditory response, candle flames were observed to respond to audible sounds.

5
Combustion of Fire–
Generated Gases

In referring to the frontispiece, thus far, we have progressed upward through the cracking and the chain reaction combustion behaviors of fuel gases and vapors. We must candidly admit that a very simplified and idealized system has been described that is never truly realized in practice. To do so would require, as covered in Chapter 1 under "Premixed Flames," close regulation of fuel flow, airflow, and intimate mixing; even under these closely controlled conditions, perfect combustion cannot be effected completely. It is this situation that is the cause of most of our air pollution problems.

PRIMARY AND SECONDARY COMBUSTION

The flames of the diffusion type (see Fig. 1.3) are the product of hostile, uncontrolled free burning that, because of the fire's exothermic nature, will grow in any direction it can, growing ever larger. This is the fire that the fire services must deal with in any of its myriad forms. A characteristic of this type of combustion is that it is *always* underventilated whether it is fully out in the open or partly enclosed. One of the reasons for this condition is the apparent existence of two phases in the overall combustion process. We might just as well, for reasons of simplicity, call the combustion that was described in Chapter 4 as the primary phase to distinguish it from what can be called the secondary phase. No new fuel, of any kind, has been added but, instead, the gaseous effluent from the primary phase passes through a temporary (very short-lived) transformation becoming alternately flammable and oxidizable. The process undergoes an alternately endothermic and exothermic series of subphases. It is these subphases that are referred to as the secondary combustion phase.

This apparent redundancy of the overall burning process can only occur when free incandescent carbon particles are present in the flame, as evidenced by its characteristic yellow orange hue resulting from the spectral emission of heated carbon. With the simple hydrocarbons, methane and ethane, the flames do not emit smoke because the incandescent carbon is virtually oxidized. Heavier fuels become progressively reddish and emit ever denser clouds of black smoke. Fuels, such as the simple alcohols, ketones, ether, and the like, that contain oxygen within their molecular structure burn with a very pale flame and are smokeless. The participants in this secondary phase are hydrogen, carbon, carbon dioxide, carbon monoxide, and steam, all of which, either singly or in combustion, emanated from the primary phase. The two phases do not exist separately but overlap to a considerable extent. Of all the oxidation reactions involving hydrogen, carbon monoxide, and carbon, the last is noticeably slowest, and this accounts for the long duration of incandescent carbon within the visible flame.

AN ANALOGY WITH SOLID FUEL-FIRED FURNACES

As was emphasized in Chapter 1, combustion when utilized for the development of useful power requires that the fuel be premixed with air in a suitable manner. This is a truism for all fuels, be they solid or pulverized, heavy residual or light and volatile, or just plain gas. The engineering textbooks are rife with information and illustrations, but for our purposes it is sufficient to refer to Figure 5.1, which illustrates the analogy of fire-generated gases as existent within a simple, old-fashioned, hand-fired, solid-fuel type of furnace. To properly aer-

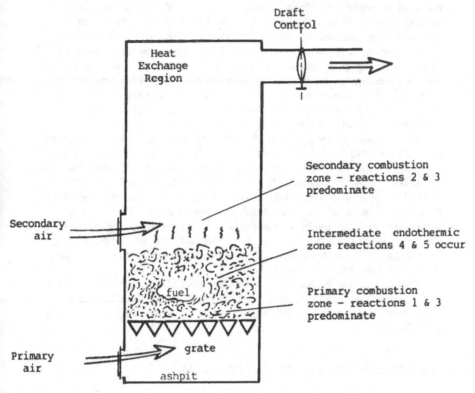

FIGURE 5.1 Analogy of fire-generated gases as existent within a solid fuel-fired furnace. Reaction (1) $C + O_2 = CO_2$; (2) $CO + \frac{1}{2} O_2 = CO_2$; (3) $H_2 + \frac{1}{2} O_2 = H_2O$; (4) $C + CO_2 = 2 CO$; (5) $C + H_2O = CO + H_2$.

ate the process requires that the fuel and air must be mixed in a controlled manner, and to do this, it is a requirement that the primary and secondary phases of combustion be separated both relative to location and air supply. For simplicity, let us assume that the fuel is coke and that the carbon and hydrogen content are the only significant constituents.

The primary air enters below the grate and then flows upward into the primary combustion zone. By pyrolytic decomposition, the coal burns according to reactions 1 and 3. The intermediate region above the primary combustion zone is where the endothermic reactions 4 and 5 occur. The heated gases, carbon monoxide and hydrogen, rise through the fuel bed and mix with the secondary air. Reactions 1, 2, and 3 are strongly exothermic and reactions 4 and 5 are endothermic, with reaction 4 being predominant (Table 5.1). Hence, the fire is self-sustaining as long as fuel and air are supplied and whatever ash is present is disposed of. The time rate of heat release is easily controllable. However, if, through negligence, the secondary air is shut off, the heated carbon monoxide and hydrogen pass through unoxidized until the secondary air supply is reestablished. This will result in a furnace explosion.

Interestingly, if the coke were to be supplied by an internal underfeed stoker instead of as shown in Figure 5.1, and no secondary air was supplied, the furnace would be converted into a crude form of an industrial gas producer. The effluent gases are rich in carbon monoxide and hydrogen, together with the atmospheric nitrogen that accompanied the air supply in the first stage of the process. They are of importance to the metallurgical and chemical industries.

The foregoing example is a vastly simplified portrayal of the distinctions between primary and secondary combustion processes and will serve as an introduction to what occurs in diffusion flames.

COMBUSTION OF FIRE-GENERATED GASES

Let us concern ourselves with the five chemical reactions shown in Table 5.1, for they represent the principal participants in the combustion of fire-generated gases. There are three elemental entities (carbon, oxygen, and hydrogen), and three compounds (carbon dioxide, carbon monoxide, and water vapor) that are derived therefrom. Their interaction is critically dependent upon the prevailing temperature levels within the flame, in which temperature gradients are very steep and reaction time is very short. If you refer to Table 5.1, it will be seen that of the six basic participants in reaction for numbers 1 through 5, the first three are exothermic and the second two are endothermic. Each reaction is thermally evaluated, in terms of kilogram calories per mole, based upon the respective heats of formation of carbon dioxide, carbon monoxide, and water vapor as derived from their elements in their standard states (18°C, 1 atm). Note that the heats of reaction for the

TABLE 5.1 Exothermic and Endothermic Reactions Involving Combustion of Fire-Generated Gases

Eq. no.	Reaction	Characteristics	Heat of reaction[a] (kgC/mol)
Exothermic			
(1)	$C + O_2 = CO_2$	Increases rapidly with temperature but non-existent above 2100°F because of Eq. (4) (see Fig. 5.2).	+94.4
(2)	$CO + \frac{1}{2}O_2 = CO_2$	Very rapid at all fire temperatures; however, CO_2 nonexistent above 2000°F [see Eq. (4)].	+68.0
(3)	$H_2 + \frac{1}{2}O_2 = H_2O$	Extremely rapid at all fire temperatures; relative reaction velocity three times that of Eq. (2)	+57.8
Endothermic			
(4)	$C + CO_2 = 2CO$[b]	Becomes significant at 1000°F; practically complete at 2100°F; combustion results in a 2:1 volume expansion (see Figs. 5.2, 5.3).	−41.6
(5)	$C + H_2O = CO + H_2$[b]	Becomes significant at 1200°F; approaches completion at about 2500°F; Combustion results in a 2:1 volume expansion (see Fig. 5.2).	−31.4

[a]Heat of formation: CO_2 = 94.4, CO = 26.4, H_2O = 57.8.
[b]Given equal temperature conditions.

exothermic type considerably exceed those for the endothermic
type. This condition is obviously a necessity for the fire to
be self-sustaining, as expressed in earlier chapters. Characteristics of the five equations are summarized in Table 5.1, and
references are made to Figures 5.2 and 5.3.

$C + O_2 = CO_2$ [Eq. (1)]

The formation of carbon resulting from the cracking process
starts with this reaction. At one time, the main use of petroleum was for the formation of kerosene and its use in lamps.
The intermediate formation of carbon was required to provide
luminescence, and control of the burning rate was necessary
for subsequent oxidation to prevent blackening of the lamp
chimney. In this day and age, carbon is a nuisance. In this
industrial age, fuels, fuel-burning equipment, and associated
controls, all are geared to the most efficient manner of combustion, with smoke being reduced to the maximum practical extent.
However, with hostile diffusion flames, huge amounts of smoke
are evolved that are as much a source of problems as the heat
released by the fire itself.

The factors that enhance the formation of soot are fuels containing

High carbon/hydrogen ratios
High fuel/air ratios
Poor mixing of fuel and air

Soot is formed in a number of stages. During the cracking
process, the progressive degradation of hydrocarbon vapors resulted in alkanes and alkenes with increasing amounts of methylene (CH_2^*) free radicals. This process formed large clusters
of carbon, together with entrapped hydrocarbons, as well as
the burning hydrogen which released hot steam. Farther up
within the flames, these aforementioned clusters break up into
small particles, a result of the combustion reaction $C + CO_2 =
2\,CO$, that are clusters of about 2- to 3-dozen carbon atoms,
which impart the yellowish color to the flame. If these small
atomic clusters are consumed within the flame, no smoke will result; however, if not enough air was supplied, these small atomic clusters coalesce into bigger particles, which are composed
principally of free carbon in unison with unburned hydrocarbons, and constitute the final smoke.

The characteristics of this reaction are that the combustion of carbon-rich particles burn at a slow rate and are everpresent in diffusion flames, particularly those from hydrocarbon and related fuels. The reaction increases rapidly with temperature: the reaction velocity is increased 1000-fold when the temperature is raised from 800°F (426°C) to 1400°F (760°C) (Fig. 5.2). The reaction is nonexistent above 2100°F (1150°C) because of the presence of the incandescent carbon itself that now reacts with the carbon dioxide it had formed (see Eq. 4, Table 5.1; and Fig. 5.2).

$CO + \frac{1}{2}O_2 = CO_2$ [Eq. (2)] and $H_2 + O_2 = H_2O$ [Eq. (3)]

The reactions typified by these equation are also strongly exothermic and are very rapid at all fire temperatures: so rapid, in fact, that they do not lend themselves conveniently to being plotted together with the other reactions. It will be seen that the carbon dioxide produced, as shown by Eq. (2), also is terminated at temperatures above 2100°F (1150°C), and for the same reason. It will also be seen that the steam produced as shown by Eq. (3) is also terminated at temperatures estimated to be approximately 2500°F (1371°C) by the everpresent incandescent carbon.

$C + CO_2 = 2CO$ [Eq. (4)]

The reaction of carbon with carbon dioxide, which is noticeably endothermic, becomes emergent at 1000°F (538°C) and is practically complete at 2000°F (1093°C). Note that the carbon monoxide produced is consumed by the oxidation reaction, Eq. (2), which in turn, produces carbon dioxide only to be reconsumed as shown in Eq. (4). This sort of "cannibalism" can only exist in the presence of incandescent carbon.

It is most important to note, when comparing Eq. (2) with Eq. (4), that in Eq. (2), 1.5 M of gases are consumed in producing 1 M of carbon dioxide but, in Eq. (4), for every mole of carbon dioxide that is consumed 2 M of carbon monoxide are produced (we are considering only gaseous moles that have volume). It is most obvious that a 2:1 pulsating volume expansion occurs that is evident to the observer. This flame behavior, which occurs only in the presence of incandescent carbon, is not to be confused with the flame flicker mentioned in the previous chapter under combustion chain reactions.

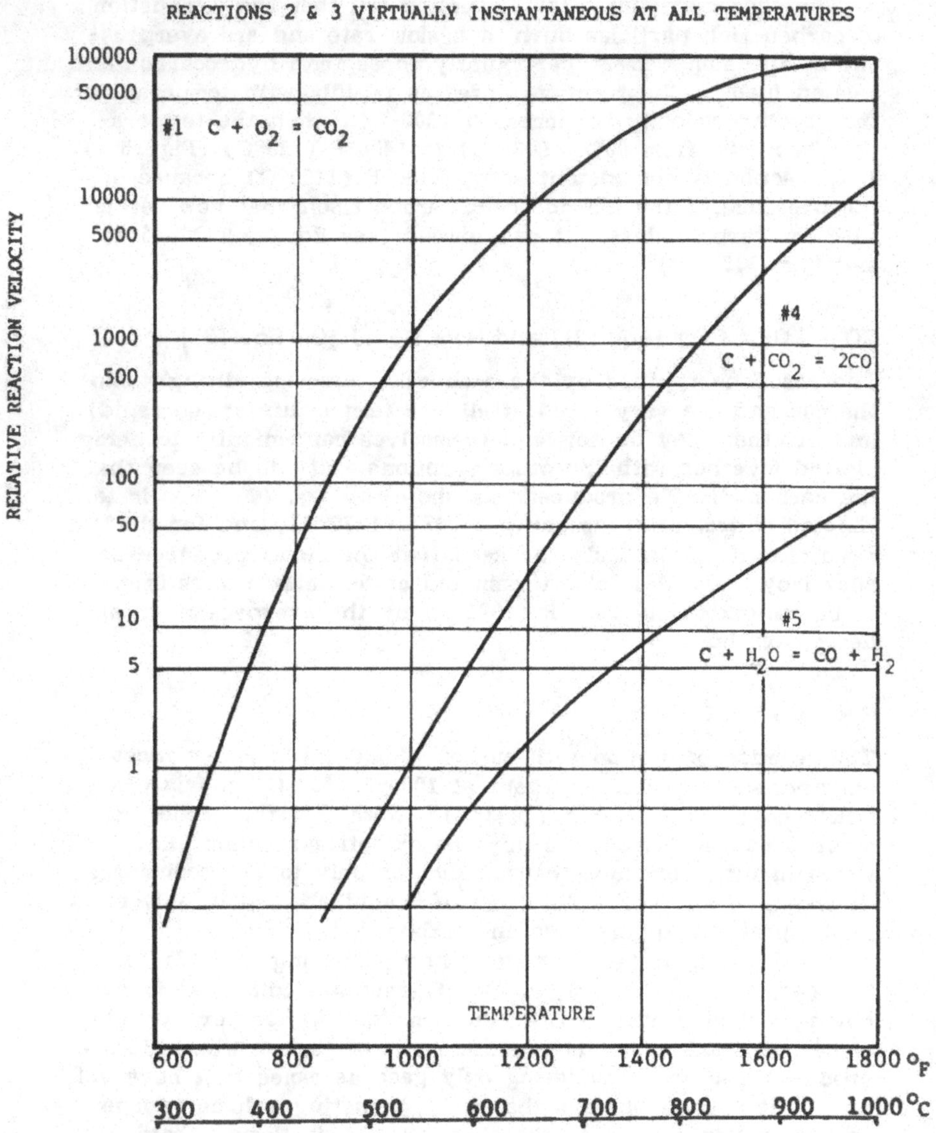

REACTIONS 2 & 3 VIRTUALLY INSTANTANEOUS AT ALL TEMPERATURES

FIGURE 5.2 Relative reaction velocity versus temperature for
reactions 1, 4, and 5 (see Table 5.1).

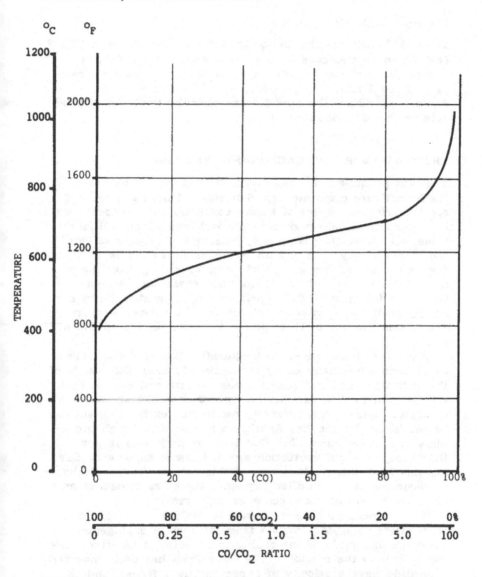

FIGURE 5.3 Composition of the equilibrium mixture of carbon monoxide and carbon dioxide in the presence of incandescent carbon. Equilibrium mixture $C + CO_2 = 2\,CO$ [from *Gas Engineers' Handbook* (1934), McGraw-Hill Book Co., New York, p. 242].

$C + H_2O = CO + H_2$ [Eq. (5)]

The endothermic reaction in Eq. (5) becomes significant at 1200°F (650°C) and approaches completion at about 2500°F (1371°C). The products of this reaction are, in turn, consumed by reactions 2 and 3 and, then, reentering into reactions 4 and 5. Again, you will see the same 2:1 expansion by the reaction products as that described under Eq. (4).

THE FLASHOVER AND BACKDRAFT SYNDROME

The twin phenomena of "flashover" follwed by the hazard of "backdraft" are commonly experienced in structural fires and are the principal causes of human casualties and economic loss. Both phenomena, although separate and occurring at different times, are, nevertheless, closely related and because of this are viewed jointly. They both relate to the flammable limits of the combustion of fire-generated gases emanating from the primary fire itself when in a semiconfined space such as would exist in a building or in fully enclosed compartments. They both oecupy positions of vital importance for fire combat strategy and tactics because of their potent influence on the spread of fire.

Flashover always precedes backdraft. Before a flashover occurs, fires are usually easily extinguished, when the fuel is of the ordinary type, and they increase in size and rate of heat release until the flames start to impinge upon the underside of a ceiling. During this interval, rescue and extinguishment can be carried on by the fire-fighting services with a high probability of success, providing that the race with time is won. Quick response and ventilation are of immense importance during this first phase. However, if this race to prevent flashover is unsuccessful, as so often happens, the extra hazard of an explosive backdraft may become an ugly reality.

After flame impingement against the ceiling, a sudden flame front starts to rapidly radiate from the area of impingement along the underside of the ceiling in an outward direction seeking an exit to the outside. The flame front has been observed to initially have a velocity of approximately 1 ft/sec, and on occasion, it has been reported to be about 10 times as much until the flames shoot out of windows and doors. Rescue becomes virtually impossible and extinguishment becomes the major task.

The reader is referred to Figure 5.4 (case B) which illustrates the flashover-prone phase that has just been described. The basis for this action is that the fire scene is semienclosed. However, if as shown in Figure 5.4 (case C), the fire scene is fully enclosed, the flashover will pose a resulting hazard that will turn into a disaster only if air is admitted, either through a door or window being opened as through a low inlet. This has been called the backdraft, or sometimes a "smoke explosion," and has all of the disastrous effects of a gas explosion. The only way that the backdraft can be avoided is for a *high* opening to be made in the structure to the outside, either by a roof vent or by forcible means.

There are a host of variables that affect these phenomena but, in general, the foregoing events take place in a very perceivable manner. Before going into some of the details and explanations, it is both timely and necessary to see that this syndrome has other aspects that are described in the following examples:

1. A project was undertaken to extinguish a residential fire by injecting huge quantities of inert gases. This was to be accomplished by directing the exhaust of a specially adjusted aircraft turbojet engine (mounted upon a trailer) into the burning dwelling. The rationale was to accomplish extinguishment by lowering the oxygen content in the fire environment. The result was tragic both for the operator, who was killed, and the house, which was destroyed by an explosion.

2. Cargo fires in the holds of merchant vessels at sea have been sealed off with the hope of smothering the fire. Inaccessibility of the cargo hampered the effective use of hose streams and steps were taken to reach port as soon as possible. The capability of marine fire protection systems is based upon prompt detection and quick extinguishing response. If the fire is not extinguished by that time, the margin for saving the vessel is greatly reduced. Sealing cargo holds is obviously ineffective.

3. The influence of building construction upon the thermal gradients at the ceiling when flame impingement occurs was forcibly demonstrated in a fire test building that had been built for the training of firefighters. The construction was massive and used reinforced concrete, with 12-in. thick walls and ceilings, and floors about 6-in. thick. The concrete contained refractory additives. The main room on the first floor had an

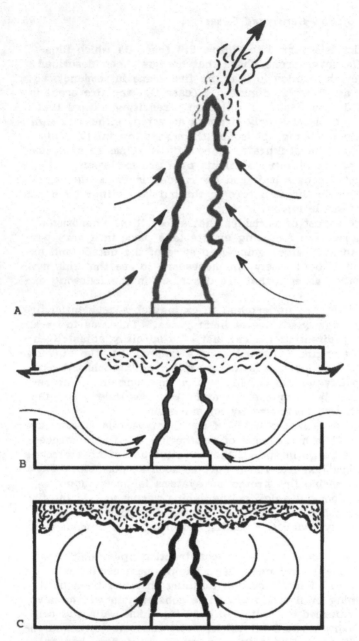

FIGURE 5.4 Effect of confinement by enclosures upon diffusion flame: (A) unenclosed and free; (B) semienclosed, *flashover* prone; (C) fully enclosed, *flashover* plus *backdraft* prone.

area of 1500 ft^2, with a 9-ft ceiling height. Window openings
had steel shutters, and doors were of fire door construction.
Hundreds of wood fires that were made from pallets and scrap
never resulted in a flashover. The maximum amount of fuel
used approximated 6000 lb (caloric value, 8000 BTU/lb). On
one occasion, 4000 lb of polyurethan foam, in the form of scrap,
was made available by a local furniture manufacturer (caloric
value, 18,000—20,000 BTU/lb). Under equivalent conditions,
such as with the previously mentioned wood fires, a pronounced
flashover occurred in 5 min. The "heat sink" effect prevented
the low-heat content wood fires from causing a flashover. The
high burning rate, as well as the heat content of the polyurethan, demonstrated forcibly that the heat sink effect had been
overcome. In this latter example, no backdraft was experienced because the sealing off of the test building was far from
perfect.

4. To show how building construction could influence flashover behavior, a test fire room was built to residential standards. It was framed with 2-in. × 4-in. lumber, faced internally with 1/2-in. gypsum board on both the ceiling and walls.
The room had a door opening and two window openings (all
were open). The test room was erected outside on the test
field. A ramp was built to the door and served as a means of
placing a wheeled steel dolly loaded with a wood crib, as described in Chapter 1 in the section on wood fires. The dolly
could be retrieved, at will, by means of a chain. At some
point after ignition, the fire, which had been slowly evolving,
suddenly erupted into a large fire with flames billowing forth
from all room openings. No backdraft occurred because the
room was open to the outside. Thereupon, the dolly with its
blazing load was withdrawn from the room. The crib fire now
subsided to its more natural unconfined appearance and burned
just as a regular outdoor bonfire. The flashover was strictly
a gas combustion phenomenon. Both the light weight and the
insulating nature of the ceiling construction resulted in high-
level ceiling temperatures [above 1000°F (540°C)].

5. Backdrafts obviously do not lend themselves to test operations. The press reports, alas, are many and fire departments are only too aware of them.

6. There are numerous incidents of coal mines that have
been on fire for years, although these are exceptional. The
general procedure is to seal off all of the available mine and
shaft openings with masonry walls or partitions in as airtight

a manner as possible, with provision made for taking gas samples for analysis. One of the examples given in Chapter 3 and shown in Figure 3.2 was taken from a mine fire. The object was to determine whether the mine atmosphere was within the limits of flammability (UFL 70%; LFL 50%). The purpose of all this is to determine how soon the mine can be reopened. In one case, 12 tons of carbon dioxide were discharged into the mine and about a month later it was decided to open the seals. The fire erupted again and the mine was resealed. About 4 months later, another attempt was made to reenter, and the fire had been extinguished. The use of high-expansion foam "plugs" blown through the mine itself has sometimes been successful. Not infrequently a mine had to be flooded and then drained to accomplish extinguishment.

The fire services have received long training on the proper techniques to be used in ventilation control. In the early stages, the spread of fire can be partly controlled by closing windows and doors, as the occupants retreat toward the exits. The intervening period, before the arrival of the fire fighters, represents the most critical period for the occupants and their safety. Once the fire fighters arrive, a decision must be made about when and how to combat the fire, remove any trapped individuals, *and* to permit controlled ventilation for the release of smoke and heated gases. If these actions take place before the backdraft occurs, the hazard and level is low but, if a flashover has occurred, ventilation becomes crucial. It must be instituted at as elevated a position as possible. The various fire-training manuals should be studied; they contain the most comprehensive treatment of the subject.

THE RATIONALE OF THE FLASHOVER/ BACKDRAFT PROCESS

The following rationale is based upon idealized conditions, which is a necessity to explain the various actions and interactions that occur. Refer to Table 5.1 and note that Eqs. (1–3) are first of all, exothermic and all three involve the straightforward oxidation of the fuel source. The equation for the reaction is declared to be exothermic when read from left to right at the completion of which heat energy has been released. To reverse this procedure would require the same amount of energy needed to return the reaction to its original form. This does

not occur in the combustion reactions but was mentioned only
to make the point that all of the equations are theoretically re-
versible. Because Eqs. (1–3), all involve a very considerable
amount of energy, in excess of the energy needed to provide
the rightward direction of Eqs. (4) and (5) the whole process
is self-sustaining, providing fuel and air are available. All
fire reactions occur simultaneously within the inferno, wherein
the products of the individual reactions act as reactants in some
other reaction as long as these reactions transpire within the
carbon-rich (yellowish) flames. Comparing the characteristics
of Eqs. (1) and (4), it will be seen that at temperatures above
2100°F (1150°C) carbon dioxide ceases to exist and, for every
molecule of carbon dioxide that disappears, two molecules of
carbon monoxide appear; a 2:1 expansion ratio of a flammable
gas, but this reaction was endothermic and required the presence of incandescent carbon. Equations (3) and (5) bear the
same relationship as described for Eqs. (1) and (4), except
that temperatures needed are in excess of 2500°F (1370°C) and
proceed at a slower pace. Figure 5.2 is a graphic portrayal of
the chemical kinetics involved for these five basic equations.
Because Eq. (4) is the primary endothermic reaction Figure 5.3
is included to related temperature with the equilibrium mixture
of the reaction

$$C + CO_2 = 2CO$$

Note that the reaction is complete at 2100°F (1150°C), but that
a drop of 700°F has rapidly reduced the CO/CO_2 ratio to a level of 5.0 and that an additional drop of 400°F further reduces
the CO/CO_2 to a very low level of 0.25. The increased formation of carbon dioxide, because of the reversibility of the reaction, will recombine with the incandescent carbon in higher-
temperature areas within the flames. The steep temperature
gradients that exist within flames account for this alternating
reversibility as well as the relative slowness of carbon oxidation.
This condition will prevail until, at the end of the combustion
process (tip of the yellowish flames), the carbon is no longer
incandescent and Eq. (1) takes over with the carbon directly
and finally uniting with oxygen to form carbon dioxide. The
carbon that is not consumed by oxygen passes off as smoke and
Eqs. (2) and (3) finally top off the combustion process by burning off residual carbon monoxide to carbon dioxide and the hy-

drogen burns off to form steam. Now, refer to Figure 5.4 and note that case A illustrates a completely free unenclosed flame, such as would be seen in outdoor fires. The reason for showing this example is to emphasize that fires start at a low level in early stages and that, by their nature, they increase in scope. Hence, although case B shows a fire of appreciable size, it is evident that when the fire was in its early stage, it strongly resembled the fire in case A but at a much lower level. Continuing with case B, the effluent of the now small fire consists of carbon dioxide, steam, and smoke that, because of the developing heat surge, rises to the ceiling, spreads out along the underside, and directs itself toward available openings. From a usually slow beginning, the flames mount and spew forth more and more effluent. The increased dynamic action of air induction into the flames creates, thereby, a circulatory motion which feeds the ever-increasing gases and smoke back into the flame. Outside air is still also being inducted. It must be remembered that all of the combustion of fire gases occurs within only the ascending visible flames. It must also be remembered that the hottest part of the flame is at its tip, where temperatures are in excess of 2500°F (1370°C). Temperatures within the flames are always lower and probably not in excess of 2000°F (1093°C).

When the flames impinge and become redirected to spread below the ceiling and the ceiling temperatures are at least 1000°F (538°C), flashover becomes imminent. Figure 5.5 illustrates the flashover "mechanism." When flashover occurs, the flammable gases pass through their lower flammable limit and consume whatever oxygen is present in the surrounding space. The mushrooming flame proceeds beneath the ceiling in all directions, accelerating as it goes toward all possible exits.

In the event that the space or compartment is virtually closed off and no more air can enter the burning area, the flames from the flashover actually become extinguished because all flammable gases have now passed beyond their upper flammable limit and the oxygen content is very low: less than 2%. All smoke and soot will by then have been oxidized, and the whole enclosed space resides at an ugly "red" heat. If an opening is made at a low level, such as a door or window, the conditions are such as to cause a backdraft or smoke explosion. Ventilation must be made at a high level, either by a smoke vent or by the use of tools. When an opening is effected the outrushing hot fire atmosphere will autoignite, and great care must be exercised in reducing the explosion hazard in a controlled manner. As one

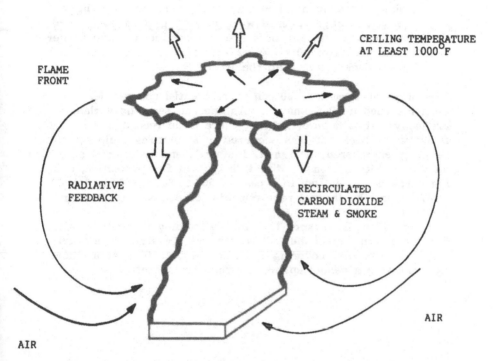

HEAT ABSORBED AND CONDUCTED
AWAY BY CEILING (NOT SHOWN)
("HEAT SINK")

CEILING TEMPERATURE
AT LEAST 1000°F

FLAME
FRONT

RADIATIVE
FEEDBACK

RECIRCULATED
CARBON DIOXIDE
STEAM & SMOKE

AIR

AIR

FIGURE 5.5 The imminent flashover.

views Figure 5.5, the question of time from ignition to flashover
cannot be answered in any definite manner. It is extremely var-
ialbe because of the relationship of the rate of heat input and
the rate of heat dissipated away through the ceiling. If these
rates are such as to allow a ceiling temperature of 1000°F or
more, then a flashover is inevitable. Whether or not a backdraft occurs is dependent on other factors that have been previously discussed.

1. The rate of heat energy *input* is dependent upon the

   a. Type and size of fuel array
   b. Arrangement of combustible surfaces

    c.   Nature of combustible surfaces
    d.   Ratio of combustible surface to room volume

2.   The rate of heat energy dissipated through the heat sink is dependent upon the ceiling construction: light-weight ceilings covered with plaster, gypsum wall board, and such, with low mass are less conductive than higher mass ceilings such as exist with reinforced concrete, steel deck plates, and the like.

Time intervals from 2 to 30 min in residential occupancies have been recorded until flashover. As one can see, the variables are many. It is noteworthy that there is no record that a flashover or backdraft has occurred in structures equipped with properly engineered, designed, installed, and maintained automatic sprinkler systems. With this type of fire protection, ceiling temperatures are not in excess of 500 to 600°F (260 to 315°C) and, hence, the thermal requirements for a flashover are non-existent.

It is also of importance that in the training of firemen, after entry has been established, hose streams are aimed in a swirling manner upward covering the whole underceiling area, thereby simulating a deluge sprinkler system in operation.

6

Fire Extinguishment

The foregoing five chapters have gone into considerable detail
about the nature of fire from its inception through the various
periods of growth. Our concerns were with the circumstances
and conditions that promoted the growth of a fire, with particular emphasis on the "hostile" variety. It became apparent that
numerous factors were involved that were not mutually exclusive. Emphasis was placed upon the exothermic nature of the
overall combustion process and the rapidity with which fire rate
and heat emission grow. Our prime objective is to arrest this
growth process as expeditiously as possible, as quickly as is
reasonably possible, and with the most efficient combination of
human and material resources. This procedure is the subject
of innumerable educational and training courses covering many
aspects of both a general and special nature.

The acts of fire extinguishment are a combination of physical skill, mental acumen, and technical knowledge and require
professionally trained people.

IDENTIFICATION OF HAZARDOUS MATERIALS

It is generally agreed that readily recognizable and easily understood markings, or symbols, on tanks and containers of all

sizes used in the transport and storage of hazardous materials, are a fundamental requirement for proper identification. The degrees and types of hazards can best be defined as embracing three basic categories

1. Health
2. Flammability
3. Reactivity (instability)

These categories are not mutually exclusive. There are two sources of health hazards: the first, arises from the inherent properties of the material itself; the second is a result of the toxic products of combustion or decomposition. Obviously, the degree of hazard will be assigned on the basis of the greater hazards. As in all exposures to toxic elements, there is a necessity to specify time limits. Exposure may vary from a few seconds to perhaps as much as an hour.

Much of this book covered the susceptibility of materials to burning. Some will burn under one set of conditions and not burn under others. The form or condition of the material, as well as its inherent properties, all affect the hazard.

The reactivity (instability) hazard concerns the degree of a material's susceptibility to release of its internal chemical-bonding energy. Some materials are capable of releasing this energy by themselves, as by self-reaction or polymerization. Some can undergo violent explosive reaction if in contact with water, other extinguishing agents, or with other materials.

For very obvious reasons the attack in emergencies that consist of any one or any combination of the three categories can be handled successfully only by trained professional fire fighters.

National Fire Protection Association Standard 704

In keeping with this logical approach and with the urging of the fire service, the National Fire Protection Association (NFPA), about 30 years ago, began the establishment of Standard 704M as a recommended procedure, and about 10 years ago this was elevated to a full standard. Figure 6.1 is a graphic portrayal of this system, and Table 6.1 is a tabulation of some of the more common hazardous materials used in our commerce. Figure 6.1 diagrammatically summarizes the full standard, which is published in the *National Fire Protection Handbook*, 16th ed. 1986,

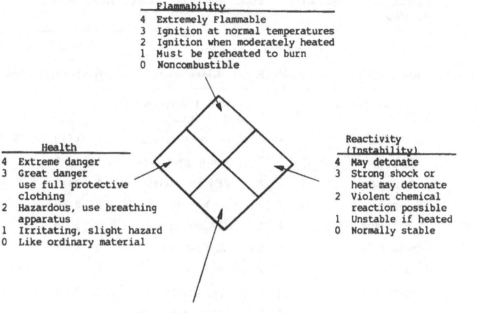

Flammability
4 Extremely Flammable
3 Ignition at normal temperatures
2 Ignition when moderately heated
1 Must be preheated to burn
0 Noncombustible

Health
4 Extreme danger
3 Great danger
  use full protective
  clothing
2 Hazardous, use breathing
  apparatus
1 Irritating, slight hazard
0 Like ordinary material

Reactivity
(Instability)
4 May detonate
3 Strong shock or
  heat may detonate
2 Violent chemical
  reaction possible
1 Unstable if heated
0 Normally stable

Special Information Symbols

Radio-
active

Use no
water

Oxidizer

Some Examples

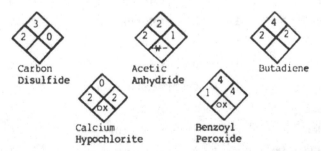

Carbon
Disulfide

Acetic
Anhydride

Butadiene

Calcium
Hypochlorite

Benzoyl
Peroxide

FIGURE 6.1 Identification system for fire hazard materials that uses a numbered diamond diagram (from National Fire Protection Standard 704).

TABLE 6.1 NFPA Classification for Some Common Hazardous Materials

Class	Rating H	F	R	Class	Rating H	F	R
Petroleum-based liquids				Common petroleum-based liquids			
methane	1	4	0	crude petroleum	1	4	0
ethane	1	4	0	motor gasolines	1	4	0
ethylene	1	4	2	JP4 jet fuels	1	3	0
propane	1	4	0	all kerosenes	0	2	0
propylene	1	4	1	JP5 jet fuels	0	2	0
butane	1	4	0	all fuel oils	0	2	0
butylene	1	4	1	all diesel fuels	0	2	0
pentane	1	4	0	transformer oils	0	1	0
hexane	1	3	0	lubricating oils	0	1	0
heptane	1	3	0	liquid asphalt			
octane	1	3	0	slow cure	0	2	0
nonane	1	3	0	rapid cure	0	3	0
decane	0	2	0	Aromatic liquids			
dodecane	0	2	0	cyclohexane	1	3	0
Alcohols				benzene	2	3	0
methyl	1	3	0	toluene	2	3	0
ethyl	1	3	0	xylene	2	3	0
propyl	1	3	0	phenol	3	2	0
butyl	1	3	0	Miscellaneous			
ethylene glycol	1	1	0	hydrogen	0	4	0
Aldehydes				acetylene	1	4	3
formaldehyde	2	4	0	ammonia	3	1	0
acetaldehyde	2	4	2	hydrogen cyanide	4	4	2
propionaldehyde	2	3	1	butadiene	2	4	2

TABLE 6.1 (Continued)

Class	Rating			Class	Rating		
	H	F	R		H	F	R
Ketones				acrylonitrile	4	3	2
acetone	1	3	0	carbon disulfide	2	3	0
methyl ethyl	1	3	0	acetic anhydride	2	2	1 -W-
methyl propyl	2	3	0	acetyl chloride	3	3	2 -W-
methyl butyl	2	3	0	acetyl peroxide	1	2	4
				calcium hypo-chlorite	2	0	2 OX
				molten sulfur	2	1	0

Source: National Fire Protection Association Standard 49.

pp. 6.9—6.13. It must be realized that, by its very nature, it is, and must be, most brief and, therefore, gives only a minimum amount of information. For instance, if water is allowed to be used in a fire situation its effectiveness in fighting flammable liquid fires is problematic if the flash points are below 100°F (38°C), and it becomes increasingly so as the flash point is decreased: hence, class I flammable I flammable liquids would be included here and would be represented in the flammability segment of the diamond-shaped diagram by the number 4. On the other hand, flammability categories 2 and 1 represent liquids within the limitations of class III, and the use of water may cause increasing "frothing" and poor fire control. These effects occur with increasing action as flash points approach and finally exceed 212°F (100°C): the boiling point of water. Hence, at the risk of being repetitious, the symbol is most useful to only trained and informed personnel.

At the very best, a hazard information system is a compromise between two conflicting requirements, i.e., immediacy of information in the event of an emergency and the adequacy of information needed for the safe transport and storage of a particular hazard(s) and, as would be expected, serious disagreements have arisen between the fire services and those personnel, such as railroad yardmasters, warehouse foremen, highway truck dispatchers, shipping clerks, and the like, who are involved in transport and storage.

United States Department of Transportation Procedure

Despite the fact that a number of large chemical manufacturers, and even certain government agencies, started to adopt the NFPA 704 system, the Department of Transportation (DOT) issued its own Hazardous Materials Regulations. All tank truck vehicles must display a designated four-digit number, located as required (on a placard or orange panel posted on the vehicle and after the letters NA or UN on the shipping papers). This identification system is based upon the system adopted for worldwide use by the United Nations Committee of Experts on the Transport of Dangerous Goods. The numbers are assigned by governmental authority under the aegis of the Economic and Social Council of the United Nations and each number has the same meaning throughout worldwide commerce. Each four-digit number identifies a specific hazardous material and has no other meaning or use.

A vital adjunct to the identification system is a manual that has been prepared to relate the identification number with a brief instruction that is intended for use by emergency personnel in the early stages of the emergency. This manual, the *Emergency Response Guidebook*, is required to be carried on each and every emergency response vehicle. Thus, upon arrival at the fire scene, after the four-number digit has been determined, the manual will have to be used to

1. Identify the chemical.
2. After that, refer to a guide number on a separate page on which is listed the potential hazards and the emergency actions to be taken.

In addition, package and container labels, which have a diamond shape but not necessarily the four-digit number, are required and are selected from a group.

The details of this system are too involved to describe here, and the reader, if he needs further information, must contact the Department of Transportation.

Certain deficiencies in current DOT placarding and labeling practices add to the difficulties in identification of the hazard-

ous nature of some commodities found in transit. Placards are
not required on the following items, either because of omissions
or because of waivers:

1. Class C explosives (pyrotechnics, squibs, small-arms
   ammunition)
2. Moderate- and low-level radioactive agents
3. Tear gases
4. Corrosive solids, including those that dissolve in water
   to produce corrosive liquids
5. "Over-the-road" equipment containing less than 1000 lb
   (454 kg) of highly toxic, flammable or oxidizing materials, corrosive liquids, flammable or nonflammable com-
   pressed gases, either singly or as a mixed shipment.

A summary of the two systems shows that a basic conflict
exists between immediacy of information and the adequacy of information chiefly needed for safe storage and transport. Certainly, the need to consult a manual on a rainy night in the
glare of headlights is not a solution to immediate identification,
nor is an arcane group of numbers of any help to shipping personnel who are not trained as fire fighters and are not privy to
the skills that are required.

THE IDENTIFICATION OF HAZARDOUS
REACTIONS

The identification systems for hazardous materials discussed so
far have been concerned with single materials only and are not
concerned with the proximity of adjacent other hazardous materials. Whether or not these various materials are mutually reactive can become a matter of great importance.

The search for, and the organization of, an extensive literature survey covering this subject was started by G. W. Jones,
of the US Bureau of Mines, who served as a member of the
NFPA Committee on Chemicals and Explosives as well as Chair-
man of the American Chemical Society Committee on Hazardous
Chemicals and Explosives. For over 30 years this activity was
officiated by the NFPA Standard 491M Committee, who published
their findings in the *Manual of Hazardous Chemical Reactions*.
It contains thousands of individual binary reactions together
with all references. It is truly a monumental work.

During the efforts that were being made to establish hazardous materials ratings for individual chemicals, concern was mounting over the inadvertent mixing of, say, at least two chemicals within a fire situation. To address this matter is extremely difficult, but if some generalities can be permitted, certain conclusions can be achieved. The US Coast Guard, together with the National Academy of Sciences, proposed a system of establishing chemical incompatability (i.e., hazard) of two generalized classes of chemicals. To keep the system within reasonable limits of simplicity, 21 separate classes of chemicals that constitute the major portion of materials shipped by land, sea, and air, and are used in commerce, were selected and are shown in Table 6.2. Of the maximum 210 possible combinations of two, 98 comprise hazardous situations that will either start a fire or contribute mightily to a fire if once started. The chemicals listed in Table 6.2 were not listed in any particular order. Although the ranking of these classes of chemicals, by themselves, is well covered by the NFPA 704 and DOT regulations, combinations of two such classes make ranking very difficult. However, if the number of incompatible reactants (as indicated by "X") are combined and retabulated in a manner such as shown in Table 6.3, certain facts become immediately apparent.

1. Organic oxides and peroxides in general are reactive with all other 20 classes of chemicals.
2. Hypochlorites, with only one exception, are also reactive.
3. Caustics, such as sodium hydroxide and potassium hydroxide, are reactive with virtually most other classes.
4. Inorganic acids, such as sulfuric and nitric acid and, to a smaller extent, hydrochloric acid, all are very dangerous from the standpoint of incompatibility.
5. In a descending order, amines and those compounds that follow in Table 6.5 can start serious problems.

Information of this type may not be of any crucial importance to fire fighters, unless they could first survey the fire area, which usually is not possible. This type of data, however, would be of great importance for fire inspectors and for storage and transportation personnel to achieve as much "separativeness," as possible, and to reduce to an absolute minimum the interincompatability of two different chemicals.

Figure 6.2 outlines the perimeters of fire extinguishment in its broadest sense taking all factors into account. Only train-

TABLE 6.2 Chemical Incompatibility Chart for Guidance in Storage and Transport[a]

CLASS

#	Class	1	2	3	4	5	6	7	8	9	10	11	12	13	14	15	16	17	18	19	20
1	Inorganic Acids																				
2	Organic Acids	x[a]																			
3	Caustics	x	x																		
4	Chlorine		x																		
5	Petroleum Base Liquids		x																		
6	Aromatic (Ring) Liquids	x	x																		
7	Alcohols, Glycols	x	x																		
8	Ketones	x	x																		
9	Aldehydes	x	x	x	x		x	x													
10	Ethers	x																			
11	Esters	x	x	x			x		x												
12	Phenols		x				x	x													
13	Nitriles	x	x	x																	
14	Amines[b]	x	x				x	x		x	x	x	x	x							
15	Org. Oxides & Peroxides	x	x	x	x	x	x	x	x	x	x	x	x	x	x						
16	Cyanohydrins	x	x	x			x							x	x						
17	Ammonia	x	x	x				x	x		x	x			x						
18	Anti-Knock Comp. (Gas.)	x				x		x				x		x	x		x				
19	Carson Disulfide	x	x											x	x						
20	Molten Sulfur				x	x									x						
21	Hypochlorites	x	x	x		x	x	x	x	x	x	x	x	x	x	x	x	x	x	x	x

[a]X indicates requirement for separateness of a pair of reactants.
[b]Example: Compatibility of amines as a class with 20 other classes.
Source: The U. S. Coast Guard and the National Academy of Sciences prepared this tabulation for guidance in storage and transport of "mixed" loads.

TABLE 6.3 Relative Probability of Any Given Class of Chemical Being Incompatible with Some Other Unspecified Class of Chemical (see Table 6.2)

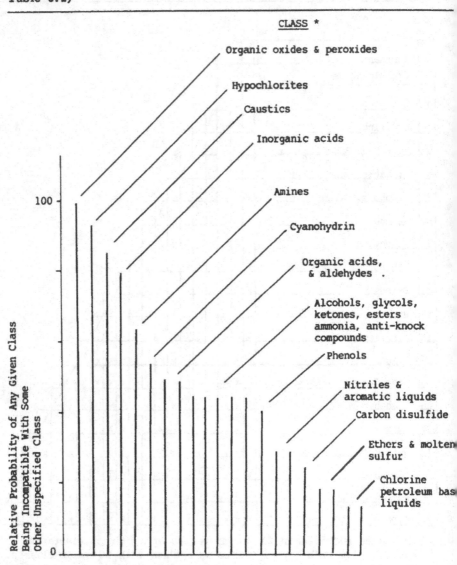

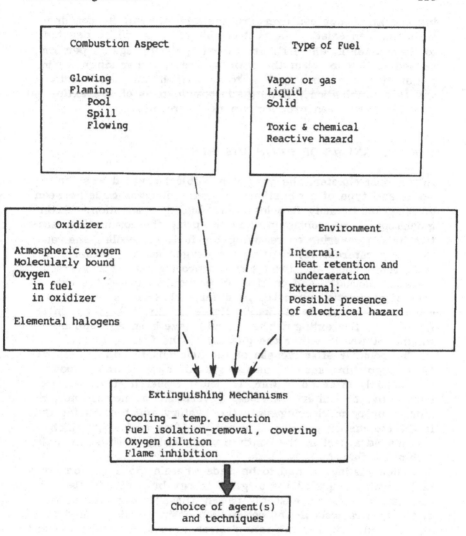

FIGURE 6.2 The variables of fire extinguishment.

ing and knowledge can overcome the virtually infinite combinations that can exist. The rest of this chapter will be devoted to the manner in which all known extinguishing agents perform, as well as how to select the optimum agent, either singly or in conjunction with other agents, to accomplish the most effective and safe combination. Synergistic combinations of fire extinguishing agents can produce remarkable results.

THE MECHANISMS OF EXTINGUISHMENT

In the first chapter, the very first subject covered was the modes and type of combustion. A distinction was made between glowing, or virtually flameless, and flaming combustion. Extinguishment, here, consisted of interrupting the combustion reaction by (1) covering or removing the fuel, (2) cooling the reaction temperatures, (3) diluting the oxygen supply, either singly or in conjunction with the other two procedures. For years, texts on chemistry and on fire, in particular, based the existence of combustion upon the need for fuel, heat, and air and called it the "triangle of fire." Hence, if this trilogy was interfered with, fire extinguishment would have been accomplished whether it was in either the glowing or the flaming mode.

The peculiar effectiveness of the halogenated and dry-chemical extinguishing agents was recognized, but not understood, until shortly after World War II. Much research effort was expended by the military services, government agencies, and private industry in clearing away the mystery and recognizing that in the combustion process intermediate "free radicals," which also played a part in the continuity of flaming combustion, exist within the flames.

A new distinction had to be made wherein glowing (nonflaming) burning required the original trilogy but, that in flaming combustion, a fourth element, namely the uninterrupted activity of the free radicals in the combustion chain reactions, had to be included. Hence, the term "tetrahedron of fire" was devised to provide for this fourth element.

Figure 6.3 illustrates the principles involved in the formation of the tetrahedron of fire. The fourth element is shown in the upper left corner of the illustration and is described in detail in Figure 6.9, and will be covered later on in this chapter. By referring back to Chapter 4 and to Figures 4.3—4.5, it will be noted the free radicals H^* and OH^* were instrumental in the

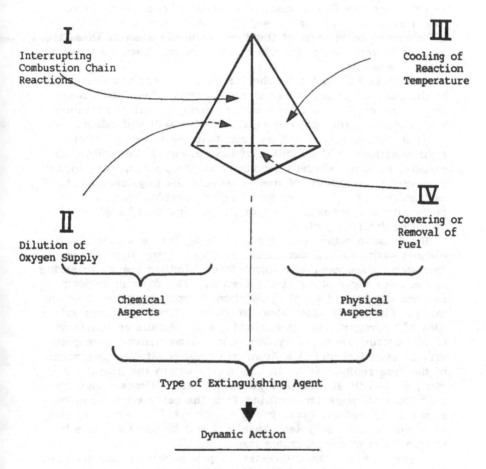

The Tetrahedron of Fire

I

Interrupting
Combustion Chain
Reactions

III

Cooling of
Reaction
Temperature

II

Dilution of
Oxygen Supply

IV

Covering or
Removal of
Fuel

Chemical
Aspects

Physical
Aspects

Type of Extinguishing Agent

Dynamic Action

Extinguisher Characteristics
Flame Velocity vs Agent Velocity
Extinguishment by Explosion (Oil & Gas Wells)

FIGURE 6.3 Fire extinguishing mechanisms.

continuity of the chain reaction and, hence, could be called
"chain carriers." The key to extinguishment is in the interruption of this continuity. In effect it amounts to a *direct* frontal
assault upon the flames wherein the heat of combustion is released regardless of the air supply, the cooling of the fuel, or
the covering or removal of the fuel. The life span of these free
radicals is very short, but their presence has been detected by
spectral analysis.

As can be inferred from the foregoing summarized description
of the chemical action, there is no lingering effect after cessa-
tion of the application of the extinguishing agent. To counter
any reignition, the operator must exercise skill and talent.

From Figure 6.3, it will be seen that the interruption of
chain reactions or the dilution of oxygen supply constitute the
chemical aspects, whereas cooling down the reaction temperature
or covering or removal of fuel constitute the physical aspects of
extinguishment. It is now necessary to combine these two aspects with the dynamic action of the application of a given type
of extinguishing agent.

By dynamic action we are not implying that a separate and
distinct extinguishing mechanism exists over and beyond the
four mechanisms mentioned heretofore. Neither are we referring
to the mere vigor of agent application. The dynamic actions
concern the rate of agent application overpowering the flame velocity. Figure 4.6 illustrates, in graphic form, the flame velocities of hydrogen, carbon monoxide, and methane as functions
of the air/fuel ratio. Very obviously, flame velocity is a property of each fuel and is a function of the relative concentration
of the free radicals (H^*, OH^*, and O^*) within the burning region, as was illustrated in Figures 4.3—4.5. Hence, flame velocity is a phenomenon resulting from the combustion chain reactions. Therefore, blowing out a methane flame would be comparatively a more easy task than it would be to blow out a hydrogen flame of a comparable size.

Figure 6.4 is a three-dimensional plot combining the effect of
flame velocity and flammability limits for a gasoline—air mixture
with various concentrations of carbon dioxide. If we refer back
to Figure 3.3, we can obtain the particular graph for methane
(which is close enough for our purposes) and transpose it on
the "XY" portion of the three-dimensional plot where flame velocity is zero. By so doing we realize that, up until now, all
test data have been taken under static conditions. The "YZ"
portion of the three-dimensional plot relates flame velocity to

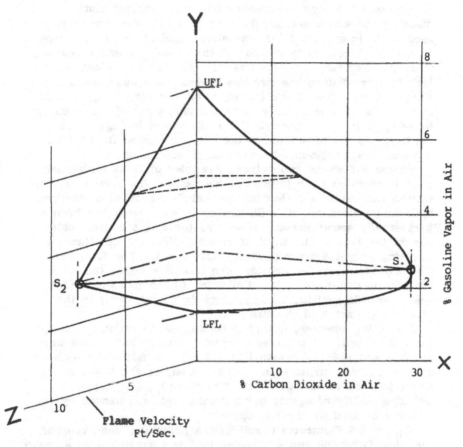

FIGURE 6.4 Combined effect of flame velocity and flammability
limits. Combustion can occur only under conditions defined
within the scope and limits of this three-dimensional plot.

gasoline vapor concentration where carbon dioxide concentration
is zero. In this situation air velocity is the only matter of importance. Any mixtures represented by points lying within the
solid volume (S_1 – LFL – S_2 – UFL) are flammable: i.e., the
application velocity of carbon dioxide can possibly accomplish
extinguishment if sufficiently high. Similarly, if no carbon dioxide is present, flame velocity by itself, if high enough, will
also cause extinguishment.

Figure 6.5 brings out another important characteristic of the discharge of a fire-extinguishing agent as it meets air resistance. Air is entrained into the stream and by an interchange of momentum results in dilution of the stream as well as slowing it down. For maximum effectiveness the range at which the attack is started upon the fire has minimum and maximum limitations. If too close, there is danger of splattering liquid fuels and, if too far away, there is insufficient vigor of the discharge to overpower flame velocity. If, as is shown in Figure 6.5, discharge is performed within the limits indicated in the form of a plume, then optimum effectiveness is accomplished.

Figure 6.6 shows two extreme examples of flame extinguishment wherein only air is used in overcoming flame velocity. Although the use of a helicopter can hardly be called a standard operating procedure, it is illustrated to show what has been experimentally accomplished. However, there is a considerable hazard involved if the angle of attack, alignment of attack, and hovering height are not properly maintained. The fire, instead of being blown off to one side, can now rise above the helicopter and be drawn down into the entry to the rotating blades. Also, for this method, there is some limit (unknown) to the size and geometry of the fire pan.

There is, however, a very practical use to which a helicopter can be put. It involves producing an artifical "down wind" for fire and rescue personnel in making a possible approach toward a burning airplane, a bus, a vessel, or the like. No attempt is made to extinguish the fire directly, but the extinguishing ability of agents being discharged is enhanced by the windage created by the helicopter.

Figure 6.7 illustrates a well-known principle, used in extinguishing "wild" oil and gas wells, that was developed by a group of specialists of worldwide fame. The preparation for making and placing an explosive bomb in proper position is a formidable task. Generally, however, it has been said that the bomb consists of a 50-gal steel drum packed with 500 lb of dynamite in a layered fashion and interspersed with a dry-chemical, fire-extinguishing agent. A remotely electronically controlled detonation device was included before sealing the drum. Figure 6.7 speaks for itself as an explanation of the obliteration of the fire.

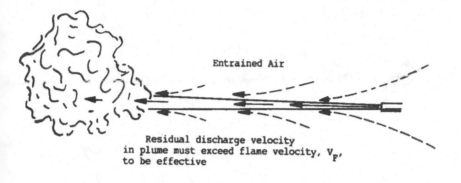

Entrained Air

Residual discharge velocity
in plume must exceed flame velocity, V_F,
to be effective

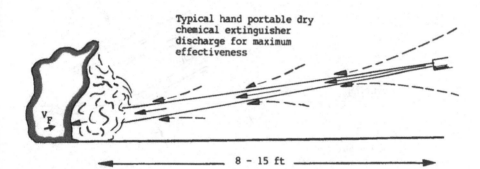

Typical hand portable dry
chemical extinguisher
discharge for maximum
effectiveness

V_F

8 – 15 ft

Typical hand portable
carbon dioxide extinguisher
discharge for maximum effectiveness

V_F

3-8 ft

FIGURE 6.5 Jet discharge of fire extinguishing agents in overcoming flame velocity as affected by air entrainment.

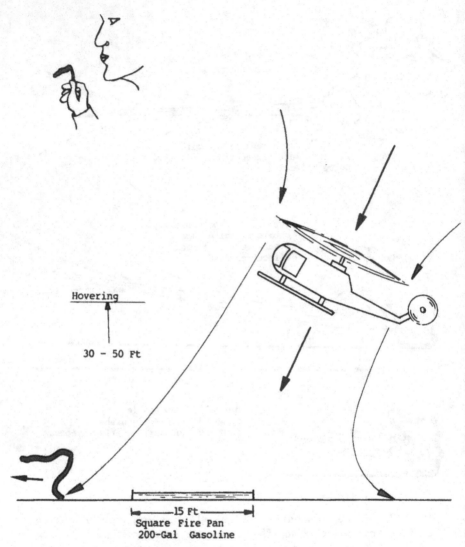

FIGURE 6.6 Two extreme examples of flame extinguishment that use only air to overcoming flame velocity.

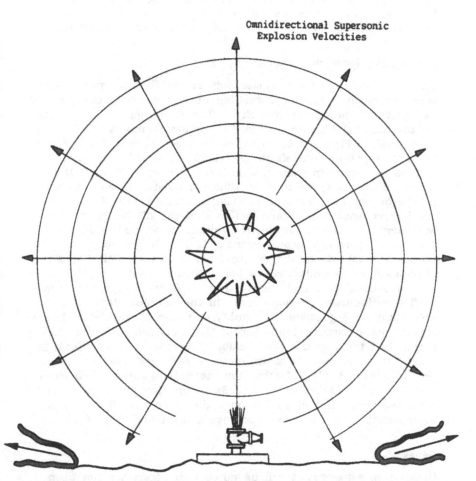

Omnidirectional Supersonic
Explosion Velocities

FIGURE 6.7 Dynamic effect of an explosion in extinguishing a
wild, oil or gas well fire.

CLASSIFICATION OF FIRE-EXTINGUISHING AGENTS

The varying chemical, physical, and dynamic natures of fire requires a classification system of extinguishing agents. The training, skill, and courage, tempered with wisdom, of the fire fighters will lead them to select the proper agent or a combination of agents. There are two systems of classification, the first of which is simplified, and the second is more detailed.

Fire Extinguisher Markings (NFPA Standard 10)

There are two aspects to classification, the first of which concerns an easily identifiable marking upon a manually operated extinguisher charged to capacity. The markings were officiated by the National Fire Protection Association Standard 10 and are illustrated in Figure 6.8. They were based upon utter simplicity and, together with simple instructions, would assist non-trained personnel in their proper use. Class A, B, and D pertain to different types of fires that can exist. Class C is *not* a type of fire but, rather, a rating granted by the Underwriters Laboratories', as a certification that the discharge stream will carry no discernible electric current when subjected to 100,000 V being impressed across a 10-in. air gap measured from the nozzle tip of the extinguisher in the direction of discharge (refer to Underwriters Laboratories', Standard 711).

The second aspect of classification defines 19 separate classes of fire-extinguishing agents that, in the current state of the art, comprise 3 gaseous, 10 liquid, and 6 solid materials. There may be some comment that some of these classes could be combined, but the way the term "class" is used, in this context, is that if the performance is different, a separate class identification is used. Table 6.4 tabulates these 19 agents. Each class is individually described, in the following discussion, along with their individual fire classification rating (see Fig. 6.8) as well as the method or methods of extinguishment (see Fig. 6.2).

Gases

Under this category, it will be noted that steam has not been included. No existing standard exists for its use. It generally is not accepted as a fire-extinguishing medium because it is ineffective unless temperatures in the space to be protected ex-

Class A:

Fires in ordinary combustible materials, i.e., wood, cloth, paper, rubber and many plastics which require cooling water (and solutions) or coating effect of certain dry chemicals.

Class B:

Fires involving flammable or combustible liquids, flammable gases which require exclusion of air, inhibiting evaporation or interrupting combustion chain reactions

Class C:

Electrically nonconductive agent to be used for operator safety in fires involving live electrical equipment. Class A and B extinguishing can be used.

Class D:

Fires with certain combustible metals, i.e., magnesium, aluminum, titanium, uranium, etc., which are combustible and on which Class A and Class B agents are *not* applicable.

FIGURE 6.8 Relationship of fire extinguisher marking to classes of fire (from NFPA Standard 10).

TABLE 6.4 Generalized Description of Fire Extinguishing Agents

Item No.		Agents	Surface fires Solid materials (A)	Non-water Miscible (B)	Water Miscible (B)	Electrical Hazards (C)	Metallic Fires (D)	Interruption of Chain reactions (I)	Dilution of Oxygen supply (II)	Cooling (III)	Covering or Coating fuel (IV)	
1	Gas (O)	Nitrogen, argon		✓	✓	✓			✓			Special Applications
2		Carbon Dioxide		✓	✓	✓			✓			
3		Halogenated Hydrocarbons		✓	✓	✓		✓				Halon 1211 Halon 1301
4	Liquid / Aqueous / Modified	Water Solid Stream	✓							✓		
5		Water Spray, Fog	✓	✓	✓	✳			✓	✓		Flash Point Above 140°
6		Detergent, AFFF, Foam	✓	✓						✓	✓	
7		Detergent, AFFF, Foam	✓	✓	✓					✓	✓	Plus a proprietary Additive
8		Diammonium Phos. Solution	✓							✓	✓	Wide Application
9		Bentonite or Borax Slurry	✓							✓	✓	Forest and Wild Fires
10		Thickened Water	✓							✓	✓	Forest and Wild Fires
11		Water plus Alkaline Salt	✓	+						✓	✓	Limited Application
12		Halogenated Hydrocarbon	✓	✓	✓	✓		✓				Halon 2402
13		Synthetic Fluids					✓				✓	Metallic Fires
14	Solid / Alkaline	Sodium Bicarbonate-Based		✓	✓	✓		✓				Standard Dry Chemical
15		Potassium Bicarbonate-Based		✓	✓	✓		✓				"Purple K" Dry Chemical
16		Potassium Chloride-Based		✓	✓	✓		✓				"Super K" Dry Chemical
17		Potassium Carbonate-Based		✓	✓	✓		✓				"Monnex" Dry Chemical
18		Monoammonium Phosphate Base	✓	✓	✓	✓		✓			✓	"ABC" Dry Chemical
19		Granular Graphitized Coke or Salt					✓				✓	"GI" or "Metyl-X"

Flammable liquid Fuels   (See Fig. 6.3)

Fire Classification (See Fig. 6.8)   Fire Extinguishing Mechanism (See Fig. 6.3)

Notes:
✳ Under Controlled Conditions
✓ Limited
+ Saponification of Fats and Grease
O Steam Not Included (No Standard)

ceed 225°F (107°C). Until the end of World War I, marine engine and boiler rooms were equipped with steam smothering systems for the dual reasons that there always was a large supply of steam and there was no other contemporary comparable agent available. Use in cargo spaces was not only useless but damaging to the cargo because of the great amount of condensation. The last ships to be so outfitted for machinery space protection were the Liberty ships and T-1 oil tankers of World War II fame.

Nitrogen and Argon

Nitrogen and argon are extremely inert and do not enter into any combustion reactions. When they are used they protect enclosed spaces only for retentive purposes. Their function is based solely upon the concept of oxygen dilution in the air, and they are suitable for fires involving class B flammable liquids. They are also more resistant to electric discharges than air itself under the same conditions. Both gases are shipped in low-pressure cylinders as cryogenic fluids as well as in high-pressure cylinders at atmospheric temperatures.

The uses for these gases are numerous but for special applications. In the field of combatting combustion the following is only a partial listing:

1. Nitrogen is used for pressurizing aircraft tires and tires on speedway racing cars.
2. Nitrogen is used to extinguishing silo fires containing coal or agricultural products.
3. Argon is used as an inert gas shield for electric arc welding and cutting.
4. Argon is used as a blanketing (not extinguishing) gas in the production of titanium, zirconium, uranium, and other oxygen-sensitive reactive metals. It is superior to nitrogen for this purpose.

Carbon Dioxide

Carbon dioxide, although not as inert as nitrogen and argon, nevertheless, is an effective fire-extinguishing agent that has been in use for over 60 years. It was been pointed in Chapter 5 (see Figs. 5.2 and 5.3) that, in the presence of incandescent carbon and at temperatures above 1000°F (538°C), the reaction $C + CO_2 = 2 CO$ occurs and is substantially complete at 2100°F

(1149°C). This reaction is primarily endothermic which, in conjunction with the large comparative rate of carbon dioxide injection, serves to overpower the flame by oxygen dilution.

In Figure 3.3 the comparative inerting ability of carbon dioxide, steam, and nitrogen is clearly shown. The superiority of carbon dioxide in rendering methane—air mixtures nonflammable has for years been ascribed to its greater molar specific heat or, in other words, its greater heat capacity. The following tabulation of molar specific heat would lend credence to this belief.

CO_2 8.8 g cal/(M · °C)

H_2O 8.6 g cal/(M · °C)

N_2 7.0 g cal/(M · °C)

The formation of carbon monoxide, mentioned previously, is also a factor that cannot be independently evaluated. Its presence can be noted, particularly when a test fire burning the usual hydrocarbon liquid fuel (heptane) is extinguished at night with carbon dioxide: the yellow flames develop the characteristic bluish fringe, and the smoke formation decreases markedly, as would be expected. A particular reason for the use of carbon dioxide is its relatively low cost. In its three available forms as gas, liquid, and solid, it is used in very large quantities by industry and, because much of this use involves food products, it is manufactured in one form only: namely, to United States Pharmacopeia (USP) standards.

National Fire Protection Association Standard 12 describes in great detail, the properties and engineering application of carbon dioxide as an extinguishing agent.

Halogenated Hydrocarbons

The last of the gaseous fire-extinguishing agents to be mentioned are the halogenated hydrocarbons. This terminology includes many compounds. Chemically they are all analogues of the simple aliphatic (chainlike) hydrocarbons in which all, or nearly all, of the hydrogen has been replaced by fluorine, chlorine, or bromine. They are more volatile and dense than their corresponding hydrocarbons and are characterized by unusually high dielectric strengths. We will confine our interest to only those of pertinence in the area of fire extinguishment. After the end of World War II, it had become obvious that of all the halogens, bromine in particular, when substituted into the molecular structure of methane and ethane resulted in an extinguishing agent of extraordinary ability. Similarly, it had also been determined that if the addition of fluorine was simultaneously accomplished in the same molecule, a much less toxic agent resulted. This had been the reason for the extended use of the highly stable Freon refrigerants. The fluorine—carbon bond linkage is the strongest to resist decomposition.

Yet, although it may sound paradoxic, a chemical extinguishing agent must decompose, to some extent, to interfere with the burning process, yet, at the same time, result in what can be called "an acceptable limit of toxicity" within a time frame.

Before delving into the fire-extinguishing mechanism of the various Halons, refer to Table 6.5, which delineates the classification system and uses a group of numerals to identify the individual halogenated hydrocarbons. It is internationally accepted.

Figure 6.9 portrays the extinguishing mechanism of Halon 1301 which is a gas under normal conditions. Exposure to the heat of the fire fractures the C-Br bond. The bromine combines with the highly reactive H^* radical that is always present in the chain reaction of combustion, as shown previously in Chapter 4 and illustrated in Figures 4.3—4.5. The hydrogen bromide formed now reacts with the highly reactive OH^* radical, which is also always present in the combustion chain reaction. Hence, H^* and OH^* are referred to as chain carriers.

The final release of free bromine frees it for reentry into the flames, thereby repeating the process. The reason for the original fracturing of the bromine atom from the Halon 1301 molecule will become apparent by referring to Figure 6.10 under the heading of individual bond energies: the C-Br bond is the "weak link" in the molecular structure of Halon 1301.

In Chapter 4, the chain reaction mechanism inherent in the combustion process and essential to the continuation of combustion was described in detail. Now, in the present example, the interference that the Halons exert upon the combustion chain process causes rapid flame extinguishment. The key to this action and counteraction lies in the removal of the H^* and OH^* chain carriers.

Halon 1301 is stored as a liquified gas in steel cylinders. It has a vapor pressure of only 200 psi at room temperature, there-

TABLE 6.5 Halon Fire-Extinguishing Agent Classification System As Originally Proposed by US Army Corps of Engineers (1950)

Number of bromine atoms

Number of chlorine atoms   per molecule

Number of fluorine atoms

Number of carbon atoms

Name of halogenated hydrocarbon	Halon classification				BP (°F)	Liquid sp gr (70°F)
Carbon tetrachloride; CCl_4	1	0	4		170	1.6
Chlorobromomethane; CH_2ClBr	1	0	1	1	151	1.9
Dibromotetrafluoroethane; $C_2F_4Br_2$	2	4	0	2	117	2.2
Dibromodifluoromethane; CF_2Br_2	1	2	0	2	76	2.3
Methyl bromide; CH_3Br	1	0	0	1	40	1.7
Bromochlorodifluoromethane; CF_2ClBr[a]	1	2	1	1	25	1.8
Bromotrifluoromethane; CF_3Br[a]	1	3	0	1	−72	1.6

[a]Liquidified gases.
Only three halogenated hydrocarbons are in general use. Halons 2402, 1211, and 1301. They offer both optimum fire extinguishing effectiveness and minimum toxic effects.

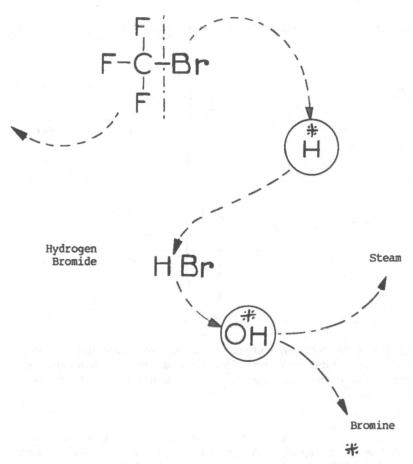

Bromotrifluoromethane
(Halon 1301)

Hydrogen Bromide

Steam

Bromine

FIGURE 6.9 Extinguishing mechanism of Halon 1301 (refer to item 3, Table 6.6).

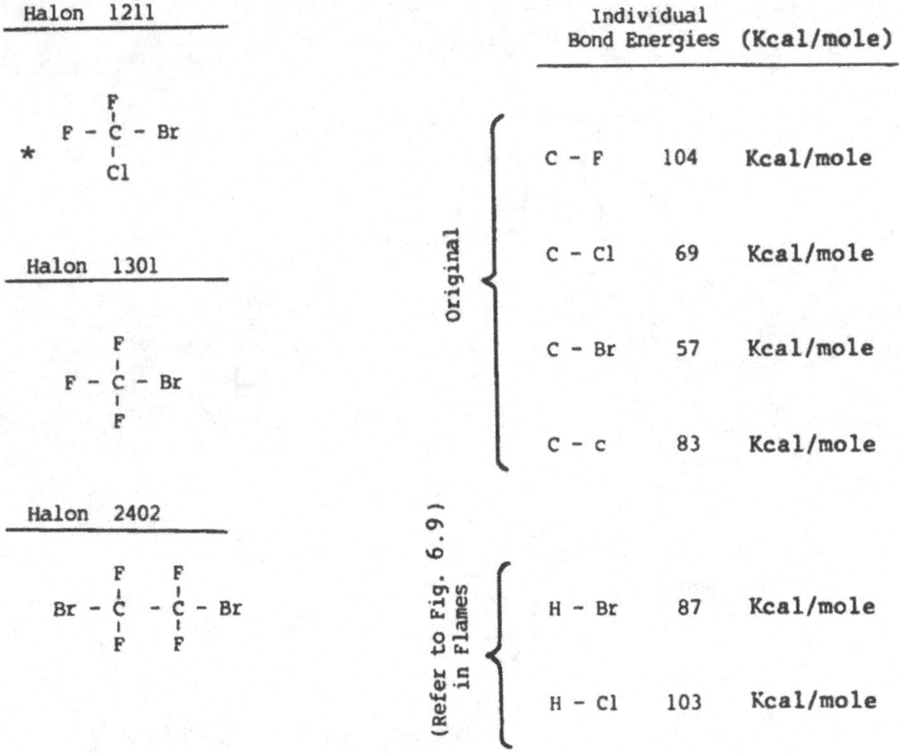

Halon 1211
_____

```
         F
         |
    F  - C  - Br
*        |
         Cl
```

Halon 1301
_____

```
         F
         |
    F  - C  - Br
         |
         F
```

Halon 2402
_____

```
         F    F
         |    |
    Br - C  - C  - Br
         |    |
         F    F
```

Individual
Bond Energies (Kcal/mole)

Original {

Bond	Energy	
C - F	104	Kcal/mole
C - Cl	69	Kcal/mole
C - Br	57	Kcal/mole
C - c	83	Kcal/mole

(Refer to Fig. 6.9)
in Flames {

Bond	Energy	
H - Br	87	Kcal/mole
H - Cl	103	Kcal/mole

FIGURE 6.10 Molecular structure of the principal Halons used in fire extinguishment, including bond energies. *Chlorine performs in a similar, but less effective manner than bromine (see Fig. 6.9).

by requiring an overpressuring (usually nitrogen) to satisfactorily expel the agent. For mechanical details the reader is referred to NFPA Standard 12A and Section 19 Chapter 2 of the 16th Edition (1986) of the *Fire Protection Handbook*, both of which are published by the National Fire Protection Association.

About 20 years ago, I conducted some experiments in a 5-ft cubic enclosure, wherein a multiple number of candles, shielded by glass hurricane chimneys, were hung at various heights. Upon introduction of gaseous carbon dioxide at a slow rate, the flames became subdued and extinguished with the glass chimneys remaining clean. However, when Halon 1301 was used instead, the candle flames disappeared in a greenish haze, and the glass chimneys were internally sooted over. The conclusion was drawn that Halon 1301 did not extinguish the incandescent carbon (as the carbon dioxide had) but, instead, by the chain reaction interruption process, stopped combustion, leaving the carbon unoxidized, in which form it was deposited on cooler surfaces.

Although Figure 6.9 concerns the extinguishing action of Halon 1301, it can be readily inferred that Halon 1211 works similarly with a small amount of hydrogen chloride being formed as well as hydrogen bromide. The main incentive for the use of Halon 1211 is that, because of its boiling point of 28°F (−2°C), it has more "liquidity" than Halon 1301, i.e., a substantially straight stream of Halon 1211 can be discharged when issuing from a cylinder having a nitrogen overcharge. It is, therefore, more adaptable for outdoor use. Halon 1301 is best suited for use as a gaseous discharge into a confined space and, in this sense, it is similar to a total-flooding application of carbon dioxide.

The advantages of using Halons 1301 and 1211 are their extreme fire-extinguishing ability, their high liquid density (sp gr 1.57 for Halon 1301, and sp gr 1.87 for Halon 1211, versus 0.76 for carbon dioxide) at 70°F (21°C), and the benefits derived from lower pressure cylinders as well as the greatly reduced weight of hardware. The main disadvantage is the relatively high unit cost when compared with relatively inexpensive carbon dioxide.

The economics influencing a choice between Halon and the carbon dioxide total-flooding systems, involves a balance between the considerably lighterweight hardware of Halon systems versus the heavyweight of the carbon dioxide hardware and the large disparity of the relative costs of the agents. There are, however, many applications for which weight and space-saving are at a premium. All modern American aircraft are equipped with Halon 1301 flooding systems for protection of engine nacelles and certain nonmanned spaces.

Liquids, Aqueous

Of the 10 generalized liquid extinguishing agents listed on Table 6.4, eight are aqueous, either in the form of water only (Nos. 4 and 5) or in a modified form, with additions used to influence surface tension, foam-forming ability, viscosity, or salts to enhance fire-extinguishing ability or to reduce the friction of flow through hoses and pipes. Water has always, and always will, provide the fire services with their first line of fire attack. Because of its high degree of latent heat of vaporization, its high specific gravity, and high specific heat, it has a considerable heat absorption capacity. Its high expansion ratio into the form of steam, although transient, nevertheless, is of value in confined spaces for producing an inert inactivation effect. Because of its general availability, quick use can be made, provided that proper equipment and trained personnel are available. When water is either not available or in limited supply, it must be transported and used in a sparing, intelligent manner.

However, despite all of the advantages, there are certain physical properties that restrict it from developing its full potential or else totally unsuitable for use on certain liquid chemicals or on burning metals. Some of these deterrents can be compensated for by modifying the water through the use of soluble additives. There are basically six restrictive physical properties.

1. Except for a municipal or proprietary water supply that is well protected, water is subject to freezing at various places and times, even in temperate climates, when contained within exposed tanks. Provided the tank is not connected to a public water supply, calcium chloride solutions containing a corrosion inhibitor will provide protection as shown in Figure 6.11. If such is not possible, auxiliary heating "winterization kits" will be required.

2. High fluid-flow friction losses are experienced when water flows through pipes and fire hose and require large expenditures of power. During the past 15 years, certain synthetic materials, which have high—molecular-weight, long-chainlike polymeric structures and are water-soluble, were found to be very instrumental in reducing the frictional drag on water flow through hose and pipes. At the relatively high flow rates used, less than 15% of this drag is produced by the viscous forces of skin friction, and the remainder is accounted for by the dissipation of energy in turbulence and vortex formation. The polymer in general use is polyethylene oxide, which is a white, granular, water-soluble material. It has been found to be effective when 1 part is dissolved with 6000 parts of water (either fresh- or seawater). The effects are startling, as in the examples shown in Figure 6.12. For a flow of 200 GPM the limits of length of 1-1/2 in. were increased from 150 ft to 600 ft

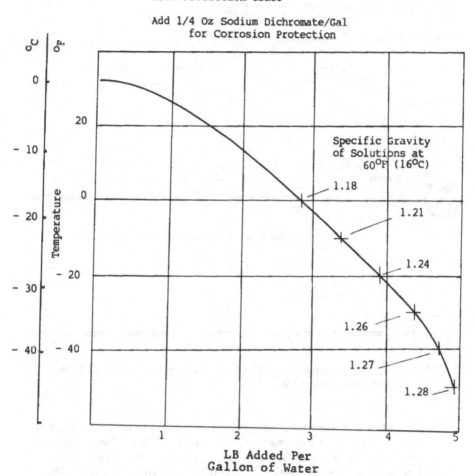

FIGURE 6.11 Freezing-point depression of aqueous solutions of calcium chloride.

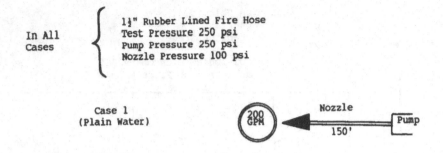

Case 1
(Plain Water)

Case 2
(Modified Water) *

* Polyethylene Oxide (1 part to 6000 parts water)

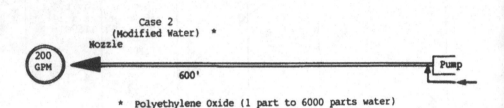

Case 3
(Plain Water)

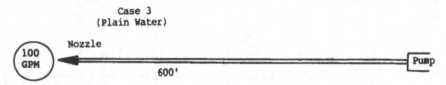

FIGURE 6.12 Comparative effect of adding a friction-reducing additive to fire hose flow.

when using the additive. By means of comparison (case 3), the limit for 600 ft of hose was reduced to 100 GPM. The numerical values used in the comparisons are approximately true. This approximation was necessary to simplify the comparison by the use of whole numbers as well as to consider hose lengths in 100-ft increments.

Consider case 2 in Figure 6.12, where it will be seen that with the injection of "rapid water" concentrate a flow of 200 GPM is achievable with a maximum length of 600 ft of 1—1/2-in. hose. In comparing the water flow rates for various lengths of hose, the pump pressure was not allowed to exceed the hose test pressure (250 psi) and the nozzle pressure was held to 100

psi. This performance is comparable with what is expected with
2—1/2-in. hose when using plain water. This characteristic is
of immense importance to firemen in hauling, lifting, and positioning charged hose lines. The *Fire Protection Handbook*
(NFPA) 16th ed., lists other experimental additions that are also white, granular, water-soluble materials, having similar molecular structures, such as polyacryliamide, polysulfonates, and
hydroxyethyl cellulose. The additive is always prepared first
as a slurry and subsequently injected either into the pump or
pump outlet hose connection. Because of deterioration, care
must always be exercised to not expose the slurry to freezing
nor to allow temperatures to exceed 120° (49°C).

3. Whereas enhanced viscosity can be a distinct advantage
in certain specific fire-fighting applications, nevertheless, it
cannot be denied that in the bulk of fire-fighting operations, a
large part of the water is lost in the form of runoff and that a
relatively small part is instrumental in actually accomplishing
extinguishment. Items 9 and 10 (see Table 6.4) will be covered
later.

4. The density of water and its lack of miscibility with liquid hydrocarbon fuels renders it ineffective unless steps are
taken to lower its density so that it can float upon the fuel surface as a foam that is relatively heat-resistant and stable.
These types of modification to water are covered in items 6 and
7 in Table 6.4.

5. The problems posed by the electrical conductivity of water cannot be altered by any type of additive. The principal
variables extend over a broad range, so much so as to render
them virtually indefinable. Therefore, the data are sparse.
The principal variables include

a. The voltage (to ground) present.
b. The purity of the water. Even pure distilled water (not
   a reality in a fire situation) is a conductor of sorts having a specific resistivity of 500,000 ohm/cm^3. Whereas,
   at the other extreme seawater has a very low resistivity
   (less than 100 ohm/cm^3.
c. The length and cross-sectional area of a straight water
   stream (if used).
d. The degree of scatter and dispersion of spray or fog.
e. The resistance to ground through the hose.
f. The resistance to ground through the fire fighter's body
   as influenced by protective clothing and footwear, skin
   moisture, and length of exposure.

6. The relatively high surface tension of water and its consequent lack of miscibility with nonpolar flammable liquids, as well as its inability to wet and penetrate various solid combustibles, present serious problems in extinguishing deep-seated glowing combustion. Chemical technology has developed a number of additives that, when mixed with water, serve to appreciably lower the surface tension and overcome the problems associated with the use of water alone and to produce results of great practical importance. These include, besides wetting and greater penetrability, emulsification, protective nonflammable aqueous films covering flammable liquids, foaming characteristics, and smaller water droplet sizes, all of which are potent factors in fire extinguishment.

The subject of surface tension is complex; however, a short description is necessary to emphasize its importance.

The molecules in the body of a liquid are subject to balanced intermolecular attractive forces of an electrical nature exerted by the molecules adjacent to them. A molecule in the surface layer, however, is subject to an unbalanced force because there are a multitude of molecules below it but very few above it. Thus, the molecules in the surface area are pulled inward and the liquid tends to assume that shape which has the smallest possible surface, i.e., a sphere. This behavior is called surface tension. It manifests itself as a contractile force acting in the surface of the liquid tending to reduce the surface area. It is as if an invisible membrane of extreme thinness existed across the liquid surface, exerting a pull within the boundary of the surface. For instance, when a camel hair brush or a group of fiber strands are immersed in water, the bristles and threads stand apart and are as free as they would be in air but, when withdrawn, they cling together by the aforementioned contractile forces of the surrounding water surface area.

Other examples of the effects of surface tension are illustrated in Figure 6.13. A sewing needle, if carefully placed upon a water surface, makes a small depression in the surface and rests without sinking despite the fact that its density is as much as eight times that of water.

The measurement of surface tension is most frequently accomplished by capillary action, as shown in Figure 6.13, using very small bore glass tubes. Because the glass tube wall is easily wetted, the liquid being tested, rises within the capillary tube above the surrounding free liquid. The word capillary is

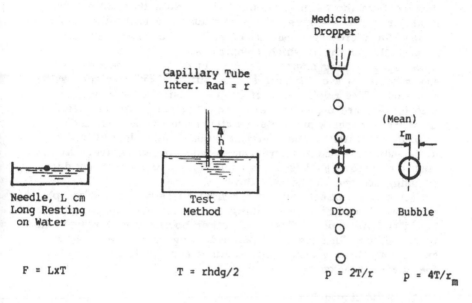

Nomenclature:

F = force, dynes
L = length, cm
r = radius, cm
h = liquid rise, cm
d = liquid density, g/cm^3
g = acceleration of gravity, cm/sec^2
p = internal pressure, $dynes/cm^2$
T = surface tension, dynes/cm

Example: Determine surface tension of acetone
using capillary tube method from following
data: r = 0.0117 cm; h = 5.12 cm; d =
0.79 g/cm^3; g = 981 cm/sec^2.

T = (0.0117)(5.12)(0.79)(981)/2

= 23.2 dynes/cm

FIGURE 6.13 Determination of surface tension.

derived from the Latin meaning "hair." Note that for a given
liquid, r and L are inversely proportioned to each other and,
hence, for microscopic values, of r, L can become very large.
Witness the height to which trees grow.

A liquid flowing slowly from the tip of a medicine dropper
emerges not as a continuous stream but, rather, as a succession
of drops. The relationship of drop (or bubble) size and surface tension are dependent upon the internal pressure. Because
these measurements and determinations are carried out in laboratories the International Metric System is used in which the
units follow the "centimeter-gram-second" nomenclature. At the
bottom of Figure 6.13 is an example for an experiment conducted using acetone as the test fluid.

Up to this moment, we have limited ourselves to surface phenomenon of surface films residing in the boundary between a
liquid and its vapor. There are other boundaries, however, in
which surface films exist. Upon referring to Figure 6.14, we
can denote three separate and distinct surface films: namely,
the surface tension of

1. Solid/liquid film
2. Solid/vapor film
3. Liquid/vapor film

If the solid/vapor surface tension is predominant, the contact angle is less than 90° and the liquid is said to "wet" the
glass. If, however, the solid/liquid surface tension is predominant the contact angle is in excess of 90° and the liquid is
said not to "wet" the glass. This is most apparent when solids
are given a water-resistant coating. There are many other examples.

The importance of surface tension and its relationship to water for the purposes of fire fighting cannot be overemphasized.
Applications will be discussed later in this section.

Plain Water (Solid Stream)

The first type of water application as shown in Table 6.4 (item
4) is the use of solid streams ejected by suitable pressure to
obtain maximum range and, at close range, to provide sufficient
impact to demolish walls, partitions, and other obstructions in
the approach to a fire.

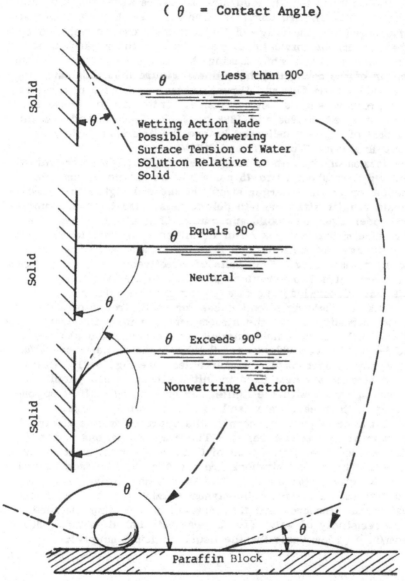

FIGURE 6.14 Influence of surface tension on wetting action.

Before World War II, solid-stream nozzles were standard equipment for the US Fire Service. Control over the volume of water discharged and the range of the streams were accomplished by the selection of nozzle tip diameter and by the regulation of fire pump pressure while making due allowance for the size and length of the connecting hose line. Figure 6.15 illustrates water solid-stream flow and thrust characteristics for nozzle size and pressure and "effective" range. Interestingly, these data were obtained in 1888 and 1889 by J. R. Freeman and presented as part of a paper delivered before the American Society of Civil Engineers (Vols. XII and XXIV).

Nozzles of sizes up to from 1/2- to 5/8-in. diameter and operating from about 50 to 75 psi would be the upper limit for manual operation. Larger-sized nozzles and higher pressures would require attachment to deluge sets, turret pipes, monitors, or ladder pipes on mobile apparatus. The effectiveness of a fire stream defies any kind of definition. Air resistance, wind conditions, and air induction into the stream cause the stream to start dispersing. The limits of effectiveness shown in Figure 6.15 are subject to variations of ± 5 ft under calm weather conditions. Crosswinds exercise strong adverse effects.

The fire-fighting effectiveness for solid streams of water becomes a reality only if the stream impinges upon the seat of the fire. With manually held nozzles, this can be, and has been, achieved for over 100 years: Solid streams of water are effective only against class A combustibles (see Fig. 6.8). Heavy solid streams are best used in initial attack, which permits a closer approach with the lighter streams at the earliest opportunity both to save water and to reduce water damage.

A matter of great concern is the approach to live electrical equipment with solid streams. The recommendations of the Hydroelectric Power Commission of Ontario, Canada, must be followed. The basic limitations are the degree of voltage hazard *and* the size of the nozzle. The Commission derived some limited data with a 5/8-in. solid-stream nozzle tip operating at 100 psi nozzle pressure, and they arrived at a limiting distance. The resistivity of water (fresh) was established as being 600 ohm/in.[3] (a low value for the usual municipal supply).

d	p	GPM	F	
1/2	50 100	53 75	19 38	
3/4	50 100	120 169	42 84	
1	50 100	212 300	75 150	Theoretical Maximum
1 1/8	50 100	268 380	94 188	

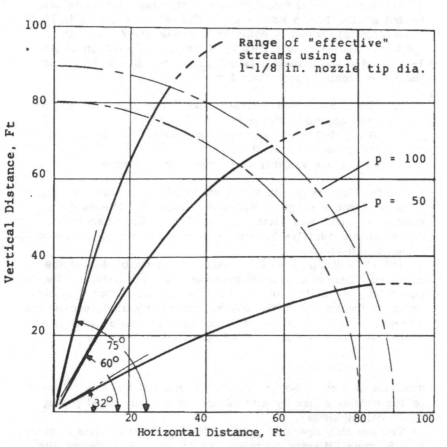

FIGURE 6.15 Solid-stream water characteristics in terms of nozzle size and pressure, flow, range, and thrust: d = nozzle tip diameter, in.; p = nozzle pressure, psi; GMP = nozzle flow; F = reaction thrust, lb; GMP = $30\,d^2\sqrt{p}$; F = $1.5\,d^2 p$.

Minimum distance (ft)	Voltage to ground (v)
15	Up to 2,400
20	4800—8,000
25	14,400—16,000
30	25,000—130,000

Note that the preceding minimum distances would all be within the range of effective streams as shown in Figure 6.15.

Another source of information was obtained from tests conducted by the Toledo Edison Co. The test was limited to 80,000 V (to ground), which is equivalent to 138,000 V line voltage (phase-to-phase). As a result, the Edison Electric Institute in 1967 adopted the following safety rules, which will limit stray leakage currents to less than 1 ma:

1. With a hand-held 1—1/2-in. solid-stream nozzle, the minimum approach distance is 20 ft.
2. With a 2—1/2-in. solid-stream nozzle, the minimum approach distance is 30 ft.
3. Do not use variably controlled spray nozzles.

Purdue University, during a study of public water supply in Indiana, found a range of 710—5400 ohm/cm^3. The hard, more-conductive water was obtained from deep wells, whereas the softer, less-conductive waters came from surface supply, such as lakes and rivers.

The wide disparity of test results is no doubt due to the combined effect of stream diameter and the variations of the volume resistivity of the water. It goes without saying that only so-called freshwater, with no additives of any kind, can be considered. Caution must be used not to extrapolate the data to other conditions because the information is too sparse.

Plain Water (Spray Fog)

The second form of water application involves the use of spray or fog nozzles either by manual operation or by fixed-piping and nozzle systems.

The manually operated type of fire apparatus nozzle is available in many different capacities and types. By twisting the nozzle tip, the water streams can be varied from the one extreme of a so-called fog discharge with a very short range that is used for drenching after extinguishment, to the other extreme of a "straightened stream." In both, the stream is in the form of small, individual jets that are discharging in a hollow, conical manner in which the conical angle can be varied from 180° to substantially 0°. The range is somewhat less than a solid-stream nozzle at the same water flow rate and nozzle pressure. Hence, the discharge can be varied from a straightened stream, to coarse sprays, to a fine mist or fog.

The effectiveness of this type of discharge lies in the fact that as the droplets of water become smaller, the ratio of the spherical (assumed) surface area to the volume of the droplet increases. The ratio becomes inversely proportional to the diameter of the water droplet. The fine misting is produced by the impingement of the individual water jets either against each other or against the flow deflector of the nozzle. A high rate of heat absorption can be achieved with fine mists; but at a sacrifice in the effective range. Therefore, the fireman will vary the nozzle setting as he approaches the fire. In any event, extinguishment is a reality *only* if the water reaches the burning surfaces.

Air entrainment will also vary with the degree of nozzle setting. It is critical that, at a nozzle pressure of 100 psi and a nozzle setting of approximately 30°, air entrainment is at an optimum level and air induction amounts to approximately 30 ft^3/min per gallon of water discharged each minute. Thus, for a water flow of 100 GPM, 3000 ft^3/min of air are inducted. Because 100 GPM is equivalent to 12 ft^3/min of water, an air induction ratio of 250:1 is obtained. This is important in the generation of medium-expansion foam. It is also important from the standpoint of ventilation, which can be either beneficial or harmful, depending upon the particular fire situation. The high degree of heat absorption that can be anticipated when using water spray or fog nozzles is shown in the upper half of Figure 6.16. The graph shows heat absorption as a function of temperature, so that if we heat 1 lb of water from 60°F to 500°F it would be proportioned as

152 BTU (12.2%), to bring to boiling from 60°F
970 BTU (77.6%), to fully evaporate
128 BTU (10.2%), to superheat steam to 500°F

Hence, every gallon of water can be expected to absorb over 10,000 BTU.

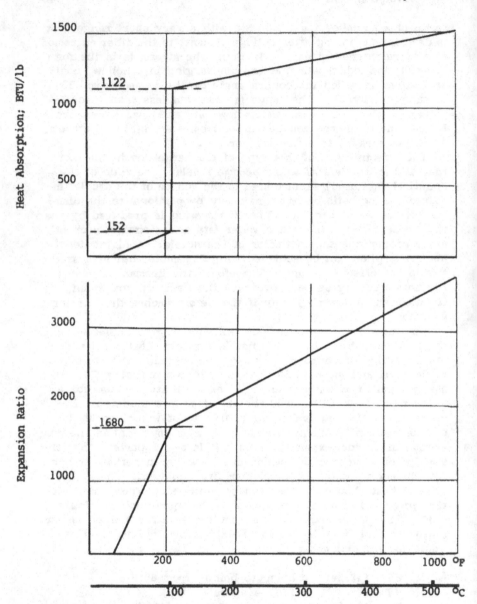

FIGURE 6.16 Heat absorption and steam expansion ratio of water based at 60°F (16°C) and heated to 1000°F (538°C).

The lower portion of Figure 6.16 shows the resulting steam formation expansion ratio. In undergoing the same temperature transition as in the preceding paragraph the expansion ratio would be about 2400:1. Because 1 gal is equivalent to 0.134 ft^3, 322 ft^3 of superheated steam (500°F) would evolve from 1 gal of water.

This expansion ratio is very instrumental in inert inactivation of the fire atmosphere, but there are three particularly important considerations that must be taken into account.

1. The expansion ratio can be considered a viable factor only so long as the temperature is sufficiently high to prevent condensation from occurring.
2. In Figure 5.2, it was pointed out that steam reacted with incandescent carbon in an appreciable manner and increasingly so at temperatures in excess of 1200°F (650°C).
3. The inert inactivating effect of steam is beneficial only within enclosed spaces and, even then, for only a limited time. Outdoor benefits are nil.

All of these constraints limit the benefits that could otherwise be achieved by oxygen dilution. Nevertheless, the powerful deterrent effects of water application on a fire are most evident, particularly so when applied in an intelligent manner.

K. Royer and F. W. Nelson, who were associated with Iowa State University, in their article, "Using water as an extinguishing agent " (*Fire Eng.*, Nov. 1963), summarized experimental activities involving small- to medium-sized room fires that were categorized as "light occupancy," the term is used by the various building code regulation committees. The concept can best be represented as a modularized space (not just a room), such as a residential occupancy, that has a total combustible content, including interior finish, trim, floor, and furnishings, of 5 lb/ft^2 of floor area, that comprises what is generally termed ordinary combustibles, e.g., wood, cloth, paper, upholstery, and such. Furthermore, the fire severity was defined as the onset of flashover. The water in spray form was efficiently applied and was carefully quantified. An empirically derived formula, which has been found to be reasonably accurate, was established and is illustrated in Figure 6.17. The accumulation of data was confined to certain ceiling heights such as would occur in conventional residential occupancies. The exam-

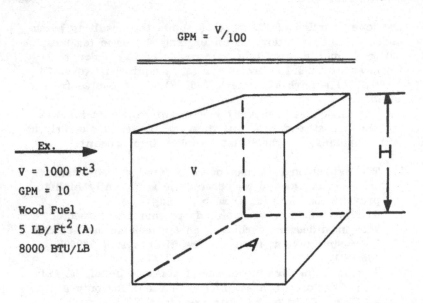

$$GPM = {}^{V}/_{100}$$

Ex.
$V = 1000$ Ft^3
GPM = 10
Wood Fuel
5 LB/Ft^2 (A)
8000 BTU/LB

H	A	GPM/Ft^2 Area
10 Ft	100 Ft^2	0.10
8 "	125 Ft^2	0.08

15 min Preburn; 1/3 Fuel Burned

Average Total Heat Release, 1,500,000 BTU

Average Rate of Heat Release, 100,000 BTU/min

Water Flow Rate, 10 GPM

Maximum Volume of Steam Produced, 3220 ft^3 at 500°F

Estimated Time of Extinguishment, 5 sec

FIGURE 6.17 Application of heat absorption and steam expansion for a residential occupancy fire.

ple shown in Figure 6.17 should not necessarily be considered as a separate room but, rather, as a modularized unit of a larger space having various ceiling heights from 8 to 10 ft. The heat of combustion of the ordinary fuels involved is taken as 8000 BTU/lb. It is apparent from Figure 6.18 that flashover would be imminent at about 15 min for the type of occupancy being considered. Usually, experience has indicated that about one-third of the total fuel load has been consumed during the 15-min interval from start to flashover. The average rate of heat release works out to be about 100,000 BTU/min. From the foregoing, 10 gal/min would be the minimum flow of water required. An extra benefit will be derived from the expansion resulting from steam formation.

The validity of the foregoing scenario and the various assumptions have been substantiated by the performance of automatic sprinkler systems that are based on the number of gallons per minute for a designated square footage of floor area. Although such systems respond more rapidly than the time consumed by the fire, as stated in Figure 6.17, the rigorous testing that sprinklers are subjected to for water flow rate versus rate of heat release is in close agreement. The reader is referred to NFPA Standard 13, "The Installation of Sprinkler Systems."

In an attempt to define fire severity, let us adopt the concept developed by the US National Bureau of Standards. This was arrived at by studying many test fires that had various temperature—time histories. With Figure 6.18 as a reference, the concept states that the area above the baseline and under the time—temperature plot of a test fire, expressed as degree minutes, is an approximate representation of the severity of a fire involving ordinary combustibles. Two fires with differing temperature histories are considered to have equivalent severity when the area under their time—temperature curves are similar. To establish a standard for comparison, it was necessary to adopt a standard, defined by NFPA Standard 251, that has also been adopted by the American Society for Testing Materials (ASTM) and the various building codes. It is used by the Underwriters Laboratories' in defining the resistance of many building components and construction, fire doors, vaults, and the like. It is illustrated in the upper half of Figure 6.18. The areas under the time—temperature curves in the lower graph are plotted as ordinates in terms of temperature (°F) times the number of minutes at various stages of the fire.

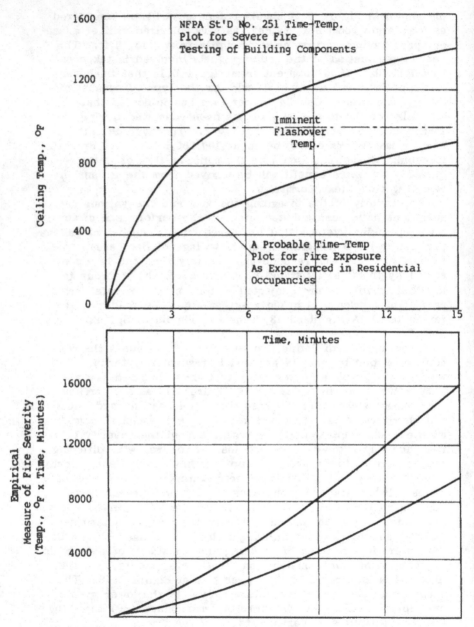

FIGURE 6.18 Time-temperature and fire severity plots (°F × min).

Finally, it must be stated that fixed and properly installed water spray nozzle systems are used for the protection of high-voltage transformers and allied equipment in electric power stations. The nozzles have special spray and range characteristics, with steps taken to ensure proper nozzle pressures. The nozzles discharge in a deluge. The systems are usually outdoors and require antifreeze provisions incorporated in the design. Minimum clearances to live electrical equipment would be as follows:

Clearance (in.)	Nominal voltage to ground (V)
12	20,000
23	40,000
37	66,000
52	93,000
63	114,000
76	132,000

for altitudes up to 3300 ft. The clearance is to be increased 1% for every increase of 330 ft in altitude (refer to NFPA Standard 15, "Water Spray Fixed Systems," for further information).

Liquids (Aqueous and Modified)

It had been stated earlier that under certain conditions pools of nonwater-soluble flammable liquid fuel could be extinguished by means of water spray streams. The extinguishing action appears to be purely mechanical, in that the velocity of the discharge can be used, in an oscillating manner, to drive flames away. The adjustment of the spray nozzle is purely judgmental. Another limiting restriction is that the discharge is ineffectual when the fuel has a flash point of less than 100°F (all class I fuels; see Fig. 2.5). Class II fuels appear to be successfully extinguished only when their flash points approach 140°F. Diesel fuel, jet fuel, kerosene, fuel oils, and the like, come within this category. The argument frequently arises that a gasoline fire was observed to be eventually extinguished. The only reason for this behavior is that in the early stages of the fire the most volatile components burn off first, leaving a residual group of components, which have flash points in excess of 140°F, that can be subsequently extinguished. In a preceding section listing the limitations imposed by the use of plain water, considerable time was devoted to the importance of surface tension and the fact that water had a relatively high surface tension value. The effectiveness of water is greatly enhanced by lowering the surface tension to achieve

1. Penetrability into fibrous materials
2. Emulsification with nonwater-soluble liquid fuels
3. The ability to form noncombustible films over liquid fuels
4. With the addition of an impurity, the ability to form foams that can float upon the surface of liquid fuels.

Modification to Reduce Surface Tension
(Nonwater-Soluble Liquid Fuels)

To achieve penetrability of solid combustibles of a fibrous nature, such as wood, cloth, bales of paper, cotton, rags, and the like, or any class A (see Fig. 6.8) combustible materials, the surface tension of the water must be lowered substantially. If we refer to Table 6.6, it will be seen that water, at 20°C, has a surface tension of 73 dynes/cm. A soap solution lowers the surface tension to 30 dynes/cm, and a usual household detergent solution will further lower the surface tension to 25 dynes/cm.

If we consider surface tension alone, we can rationalize how water, for instance, can penetrate into the minute interstices that exist in all fibrous or laminated materials, when the surface tension is lowered enough. This action is called "wetting." The further the surface tension is lowered, the greater will be the penetrability. Penetrability is a vital function in fire fighting to enable the water to reach hidden glowing combustion that, if not extinguished, can reignite a fire that had only been superficially extinguished.

The term "detergent," (derived from Latin *detergere*, to wipe off) applies to those additives (either liquid or solid) that upon dissolving in water reduce its surface tension. But, in addition it imparts the very important property of "emulsifying" the water with a nonwater-soluble liquid flammable fluid. The emulsion formed is nonflammable. The emulsifying agent is called a surfactant. It is best defined by referring to the top of Figure 6.19 which indicates that the molecule consists of two

TABLE 6.6 Relative Surface Tension Values for Water and Some of Its Solutions with a Variety of Liquid Flammable Fuels

Nonaqueous Liquid	Surface Tension (dynes/cm)	Aqueous Liquids
		73·(20°C) ⎫
	70	70·(40°C) ⎬ Fresh Water
		66·(60°C) ⎬
	60	62·(80°C) ⎭
48 Ethylene Glycol	50	
44 Nitrobenzene		
		45 Protein Foam Solution
	40	
29 Benzene		
28 Toluene		
25 Methyl Alcohol[a]	30	30 Average Soap Solution
23 Acetone[a]		
22 Ethyl Alcohol[a]		25 Av. Detergent Solution
22 Octane		
21 Acetaldehyde[a]		
20 Heptane		
18 Hexane	20	
17 Ethyl Ether[a]		
14 Pentane		16 AFFF Solution
	10	
	0	

(Vertical axis label: Surface Tension Dynes/CM)

[a]Naturally water-soluble.

Surfactant
Molecule

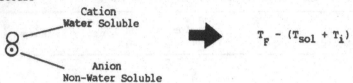

Cation
Water Soluble

$$T_F - (T_{sol} + T_i)$$

Anion
Non-Water Soluble

Must have a positive value
for film to spread

Where

T_F = surface tension of fuel

T_{sol} = surface tension of AFFF water solution

T_i = interfacial tension between
fuel and solution (1-5 dynes/cm)

Retreating
flame front

Omnidirectional
advancing film

T_{sol}

T_F

T_i Non-Water-Soluble
Liquid Fuel

FIGURE 6.19 Influence of surface tensions upon formation of
sealing film on a liquid fuel surface.

parts: a water-soluble *cation* and a nonwater-soluble *anion*. The cation dissolves in water and the anion dissolves in the nonwater-soluble liquid fuel, thereby forming an emulsion that is nonflammable. On the left side of Table 6.6 are a group of 13 chemicals: seven are nonwater-soluble, and six are soluble. The numbers on the left side of these chemicals indicate their respective surface tensions. They indicate how much the surface tension of water must be lowered to accomplish the emulsification "unions." Thus, for benzine (C_6H_6) ordinary detergents can lower the surface tension to that required for emulsification (29 dynes/cm or lower). There were limitations to this procedure until about 30 years ago when chemical technology developed new and improved detergents that by a unique method of action could form sealing films over liquid flammable fuels that were nonwater-soluble. These films became known as AFFF, short for aqueous film-forming foam.

Detergents have now been developed that lower the surface tension of water to levels that previously were never achieved. The upper part of Figure 6.20 illustrates the structural formula of alkyl perfluorosulfonamide, which comprises two main groups: the long alkyl chain, in which the hydrogen atoms have been replaced by fluorine atoms, is the oil-miscible portion of the molecule; the second portion is miscible in water and polar solvents. The first group is termed hydrophobic, a property characteristic of hydrocarbon chain-type molecules. The second group is termed hydrophilic; it is highly soluble in water because of the presence of hydrophilic groups in solution, such as OH^*, CO_2Na^*, SO_3K^*, and amine NH^*. A surfactant's surface activity depends upon the sizes of these groups and upon their relative solubility. Because of the differing solubilities, surfactants are surface active, i.e., they accumulate at surfaces. In particular, the hydrophobic groups accumulate at an air-water interface where the hydrophilic groups also gather. Thus, emulsification takes place in the form of a film, thereby preventing fuel vapors evaporating to feed the fire.

From Figure 6.20, it can be seen that for the film to spread and cover the nonwater-soluble liquid fuel surface, certain conditions will have to be met. The combined surface tension of the AFFF—water solution (16 dynes/cm from Table 6.6) and the interfacial tension that exists between the AFFF—water solution *and* the nonwater-soluble liquid fuel (which varies from 1—5 dynes/cm) must total *less* than the surface tension of the fuel

"AFFF"
A Derivative of
Alkyl Perfluoro – Sulfonamide

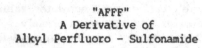

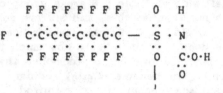

Hydrophobic anionic
non-water miscible
portion

Hydrophylic cationic
water miscible

Type 1

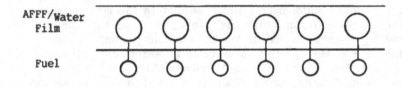

AFFF/Water
Film

Fuel

Type 2

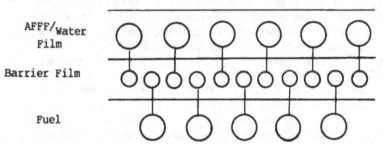

AFFF/Water
Film

Barrier Film

A Cohesive
Polymeric
Additive

Fuel

FIGURE 6.20 The sealing of burning flammable liquid surfaces of nonwater-miscible and water-miscible (polar) fuels: Type 1, nonwater-miscible hydrocarbon; type 2, polar solvent.

itself for the film to spread in an omnidirectional manner across
the surface of the fuel. Note that the aqueous film floats on
top of the lower-density liquid fuel in spite of being heavier. A
similarity exists when compared with the needle resting on plain
water (see Fig. 6.13). Therefore, from the left side of Table
6.6, we see that octane and heptane (and their isomers) represent the lower limit at which AFFF—water solutions can be expected to perform. Hexane and pentane are beyond the scope.
Gasoline comprises hydrocarbons ranging from the C_5 to C_{12}
groupings, thus, pentane and hexane must selectively burn off
first, after which the higher carbon groupings are covered with
the film preparatory to extinguishment.

To apply AFFF—water solutions in an effective manner, they
should be dispensed as a spray without unduly agitating the
liquid fuel surface. The AFFF—liquid concentrate must be mixed
with water according to the manufacturer's recommendations.

One such source, for instance, has a product that is recommended as a 3% mixture for nonwater-soluble fuels and a 6% mixture for water-soluble fuels. The following limited test results
are significant. The tests were conducted according to Underwriters' Laboratories standards using a 50-ft^2 square pan with
a 2-in. fuel depth and a 1-min fire-burn interval.

Concentration	3%	6%
GPM/ft^2	0.04	0.10
Fuels used	Gasoline	Methyl alcohol
	Heptane	Ethyl alcohol
	Toluene	Acetone
Extinguishing times (min)	2 to 2-1/2	2 to 2-1/2

An accompanying foam was developed having an expansion
ratio that varied from 6:1 to 8:1.

As a preliminary to the subject of fire-fighting foams it is
suggested that the reader obtain the following NFPA Standards:

11   Foam Extinguishing System
11A High Expansion Foam Systems
11B Synthetic Foam and Combined Agent Systems

and study them to obtain the information they desire on this
most complex subject. It is also suggested that the text: R. L.
Tuve (1976). *Principles of Fire Protection Chemistry*. National
Fire Protection Association, be consulted on this same subject.
Dr. Tuve and his colleagues were directly responsible for much
of this development.

As stated at the beginning of this text, it was not the intent
to merely repeat what has already been amply covered to a
large extent and in considerable detail by the technical literature. The total subject cannot be satisfactorily covered in this
text. An excellent coverage of the use of fire-fighting foam is
also contained in the *Fire Protection Handbook*, previously mentioned in this text.

Because foams are lower in density than the aqueous solutions from which they were generated and are lighter than the
flammable liquid fuels (nonwater-soluble) upon which they float,
the combined effect is to exclude air, promote cooling, and provide a continuous layer of vapor sealing either for preventing
combustion or for protecting the fuel from burning.

The ability of a water-based liquid to become a frothy mass
of bubbles requires a balance of forces and conditions not entirely understood. Foams are unstable "air—water emulsions"
and are sensitive to physical or mechanical forces. Exposure
of the foam blanket to intense radiant heat causes it to break
down because of evaporation of its water. The only way this
effect can be minimized is to supply the foam at a substantially
higher rate than it is being depleted. Foam formation is also
sensitive to various chemical vapors, as well as the gaseous
products of combustion. The manufacturer's instructions must
be closely followed.

Two factors govern the ability of a liquid to froth into a
foam: the first is surface tension; the second, which is para-
mount in creating a foam, is the presence of an impurity. A
pure liquid cannot form a stable foam in the sense that after
vigorous stirring, any residual froth remains. Foamability depends upon a surfactant as well as other impurities added to
perform specific tasks. These intricate factors have unforseen
remarkable consequences. For example, all beer drinkers must
have observed that in the moments after beer is poured into a
glass, the volume of the "head" shrinks while that of the liquid
grows. The bubbles are roughly spherical. Yet few beer
drinkers know that as the foam drains, increasing the volume

of liquid beer, the bubbles remaining in the foam respond to a
changing balance of forces by changing from spheres into poly-
hedrons.

In real foams the basic distinction is between "wet" foams and
"dry" ones. In a wet foam, the liquid content is high and the
walls are consequently thick enough that the bubbles are not
distorted. Hence, each bubble, independently of the others,
minimizes its surface area by taking a spherical shape. As the
foam drains its geometry becomes more complex. The cells, now
separated by thinner walls, begin to influence one another and
change from spherical to polyhedral shapes, as described by
the Belgian physicist Plateau over a hundred years ago. Each
film is a smooth surface. The films are flat only if the pressures in two adjacent bubbles are equal. Moreover, the films
meet only in sets of three at angles of 120°. Figure 6.21 illustrates that when bubbles of unequal size join and coalesce, the
films are distorted. Whereas bubbles of equal size have a flat
common film, and pressures within the bubbles are equal. The
optimum condition, therefore, exists when an isometric cellular
matrix is formed that has a slow decay and, hence, the highest
longevity. The matrix is remarkably resilient and will not interfere with the need for the foam to flow.

Foams, at first (circa 1903), were produced by the reaction
of two liquid solutions, one alkaline (sodium becarbonate plus a
foam stabilizer) and the other acidic (aluminum sulfate). The
reaction generated copious quantities of carbon dioxide gas, resulting in a foam having eight times the total volume of the individual liquids. These foams consequently had a high mineral
content of sodium sulfate and insoluble aluminum hydroxide
which stiffened the foam bubbles to the extent that under radiation from a fire, fluidity of the foam was impaired during dehydration and subsequently would "cake," breaking the foam
blanket. The foam stabilizer used with the alkaline solution was
usually hydrolyzed vegetable protein. In fact, one of the first
stabilizers used was a hydrolyzed extract from licorice root.
The foam, so created, has become popularly known as "chemical
foam" and was adapted for use ranging from small hand-portable
units to large systems both mobile and fixed. In some units,
solutions were separately and initially prepared and stored for
use on a yearly basis. In other instances, particularly in mobile equipment, the two individual dry chemicals were stored in
a mixed and dry condition in containers and used on demand by

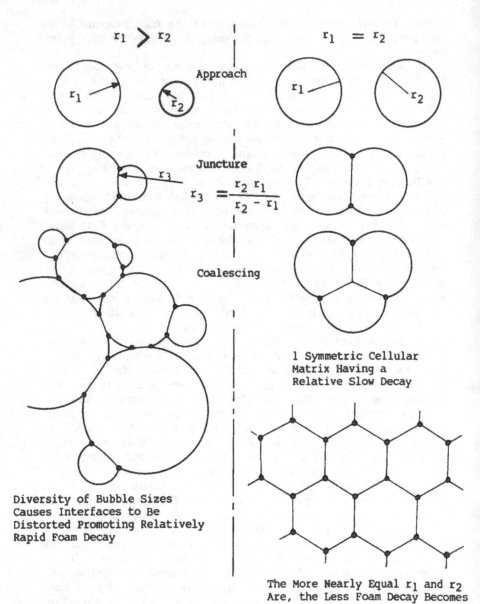

$r_1 > r_2$ $r_1 = r_2$

Approach

Juncture

$$r_3 = \frac{r_2 \, r_1}{r_2 - r_1}$$

Coalescing

1 Symmetric Cellular
Matrix Having a
Relative Slow Decay

Diversity of Bubble Sizes
Causes Interfaces to Be
Distorted Promoting Relatively
Rapid Foam Decay

The More Nearly Equal r_1 and r_2
Are, the Less Foam Decay Becomes

FIGURE 6.21 Influence of relative adjacent foam bubble sizes
upon foam decay.

means of a hopper feed into a moving stream of water, thereby making the foam in transit through hose or piping to the fire site.

Note that this type of foam does not depend upon surface tension to assist in its production. The foaming agent (carbon dioxide) is trapped within the bubbles and was formed in a so-called internal manner as contrasted with other fire-fighting foams for which the frothing is produced in a so-called external manner by means of air induction. Chemical foam was the first means of controlling large-scale flammable liquid fires and has become surpassed by other fire-fighting foams from the standpoints of economics and maintenance difficulties.

During the interval between World Wars I and II, mechanical foam was developed which used air as the frothing agent. In this scheme, water and a liquid foam stabilizer are mixed into a common solution and subsequently discharged through an air-induction aspirator in which, by various mechanical means, a strong scrubbing of the air and the solution takes place, thereby creating a foam.

Essentially there are two basic types of foam stabilizers with each type having two subtypes.

Protein based		Detergent based	
Plain protein	Fluoro-protein	Aryl and alkyl sulfonates	AFFF
(1)	(2)	(3)	(4)

1. The protein base of the stabilizer is the product of a thermal degradation by hydrolysis of protein-based materials, such as blood, horn, and hoof slaughter house products; fish scales; and various vegetable materials such as soy beans, cotton seed, peanut shell and similar meals after oil extraction. During the cooling process water is absorbed and amino acids and peptones are formed. To this mixture, mineral salts, that can be readily oxidized are added to impart fire resistance. Finally an emulsifying agent is added to assist in dispersing air into the fluid to the maximum degree. The mineral salts are instrumental in providing the foam bubbles with a skeletal structure during the dehydration that occurs when the foam is subjected to the radiation of the fire. Protein-type concentrates are used in either 3 or 6%, by volume, fresh- or seawater.

Usually, these concentrates produce dense viscous foams of good
stability and better resistance to "burnback" than most other
foams. They are sensitive to air exposure and form scums.
They must not be premixed with water, for then the hydrolysis
reactions would proceed once more resulting in a worthless stabilizer. They are sensitive to exposure from dry chemicals if
the latter are used simultaneously upon a fire. The sensitivity,
as evidenced by foam breakdown, is variable.

2. In 1965 an improvement was made by fluorinating the protein-based foam stabilizer. The hydrolyzed protein component
is protected by the presence of a partially fluorinated agent
that is loosely bonded thereto. Compatibility with the use of
dry chemicals in fire situations is improved. The foam is more
stable and less affected by exposure to air. Subsurface injection of foam streams into storage tanks is possible. These
foams possess superb reflash resistance and are a marked improvement over foam (1). These concentrates are available in
3 and 6% concentration grades.

3. Detergent-based foam stabilizers are essentially a blend
of aryl and alkyl sulfonates. They can be blended with water
for indefinite storage periods and are compatible with the use
of all the varieties of fire-fighting dry chemicals. However,
they do not have the same degree of fire resistance to decay
nor the better resistance to "burnback" that protein-based foams
have.

4. Aqueous film-forming foam (AFFF) belongs in the detergent-based grouping. However, it is characterized by its ab-
normally low surface tensions compared with other foaming mixtures, as shown in Table 6.6. When compared with the protein-based foams (1) and (2), it resembles foam (3) in having
lower heat resistance. However, considerable research has established solution application rates that definitely show that
equal fire intensities could be countered with solution rates of
half those needed with protein-based foams.

When reviewing the various types of foams and their merits
and weaknesses, it becomes evident that the NFPA Standard 10
and the manufacturer's instructions must be read and observed
in a very literal manner. When correctly used, any of these
types of foams do an excellent job in combating flammable liquid
fuel fires. In fact, for very large hazards, such as exist at
airports, oil refineries, many chemical plants, and the like,
foam is the only practical medium that can be used. At the conclusion of World War II, a US Admiral remarked that the reason

our Navy could remain afloat and fighting was, among other factors, because of foam-type fire protection. Every air force
base, every major commercial airport, and every major refinery
has extensive foam systems and oil and fuel storage tank farms.

The rapidity of flame extinguishment is, invariably, of a low
order, and it becomes a matter of extreme importance to attack
the fire as soon as possible. Supplementary auxiliary fire equipment must be pressed into action, with care not to harm the
foam by the excessive use of water or the use of noncompatible
dry chemicals. In the form of sprays or fogs water can, however, be of great assistance.

Last, but most certainly not least, is the contribution that
the use of type (3) foam (detergent-based) made in developing
the use of "high-expansion" foams (expansion ratios from 100:1
to 1000:1). Their use is covered specifically in the NFPA
Standard 11A. They are especially suited for class A and class
B fires in confined or semiconfined spaces that are particularly
suited for total flooding. The use of this type of foam outdoors
is limited because of its low density, which makes it vulnerable
to be blown away by winds or other adverse conditions. Table
6.7 is a compendium of the whole range of foam expansion ratios, foam density, and volumetric water content. Added to all
of the advantages of using water for heat absorption and steam
formation, are the advantages of low surface tension provided
by the penetrability of burning fibrous solid fuels. It should
be no surprise that the decay of foam can be accelerated by exposure to the radiant heat of the fire. This can be countered
by applying the foam/water solution at a rate high enough to
provide a sufficient quantity of the foam. The more gently the
foam is applied, the more rapid the extinguishment and the lower the total amount of agent required. Experience has demonstrated that the high- and medium-expansion foams can be accumulated in depth and provide an insulating barrier for protection of exposed flammable materials and structures adjacent to
the burning area. Furthermore, by referring to Table 6.7 it
will be apparent that at elevated expansion ratios, the semista-
bility of the foam and the consequent lowering of foam density
causes the water to be temporarily "captured" in depth so that
when urged to flow by means of supplied ventilation, it will be
transported to all spaces both accessible and inaccessible. This
feature prompted the development for use of high-expansion
foams in combating coal mine fires by the Safety in Mines Research Establishment in Buxton, England. It was found, for in-

TABLE 6.7 Foam Expansion Ratios, Density, and Volumetric Water Content

Degree of Expansion	Foam Expansion Ratio	Specific Gravity	LB H_2O/Cu.Ft.	Cu.Ft./LB H_2O	Cu.Ft./Gal. H_2O
High (Confined or semi-confined spaces only)	1000	0.0010	0.063	16.0	133
	800	0.0013	0.078	12.8	106
	600	0.0017	0.104	9.6	81
	400	0.0025	0.156	6.4	53
Medium	200	0.0050	0.313	3.2	27
	100	0.0100	0.625	1.6	13
Low	10	0.1000	6.25	0.16	1.3
	5	0.2000	12.50	0.08	0.67
	0	1.0000	62.50	0	0

Increasing Density, Fluidity, and Wetness →

stance, that a 1000:1 expansion ratio foam could be forced through mine tunnels and shafts for considerable distances in advance of the foam-generating apparatus. The amount of water used was relatively small, thereby not causing water flooding to occur.

Air foams are designed to be stable when generated with either fresh- or seawater, at normal ambient temperatures from 35 to 80°F (2 to 27°C) and free from contaminants such as chemical plant discharges, oil residues, and sewage pollution. Foams are also adversely affected by air that contains various gaseous combustion products. Although this latter circumstance is usually not an important factor with normal foam application, it is necessary to locate fixed foam makers on the sides of, and not above the hazard.

The plethora of all the various types of storage, proportioning, and piping systems, the virtually endless installation applications, and the large array of dispersing means is too large to be encompassed in this text. Detailed technical advice is available from the various approved agencies, technical societies, and manufacturers' bulletins. The strengths and sensitivities of foam application require strict observance of the manufacturer's instructions for prompt and successful extinguishment.

It is most important to relatively proportion the admixture of water and liquid foam concentrate and, although there are numerous ways that this can be accomplished, there are two general methods in use.

1. By utilizing the energy of the moving water stream by means of Venturi devices or orifices to naturally induct the liquid foam concentrate.
2. By using auxiliary pumps and not relying on the water flow energy.

The *Fire Protection Handbook* (NFPA) lists 10 different means of proportioning in general use. Some are meant for rather simpler application by portable equipment, whereas others are more complex and meant to be included in fixed-piping systems, usually of large scale.

Figure 6.22 illustrates the three basic stages required for the generation of fire-fighting foam of all *kinds*. The method most frequently used by the fire service and also in general use combines the three basic stages into a single piece of equipment. The reader is no doubt wondering why there is such a variety of foam compounds and foam-generating apparatus. The answer lies in the course of development for three-quarters of a century, coupled with the multiplicity of applications and a rapidly expanding petrochemical industry. It is not too extreme to predict that this large and complicated array of equipment and materials will be considerably condensed and simplified with concentration upon the synthetic high-expansion and the aqueous film-forming foams (AFFF): both of these categories would protect against external as well as internal hazards; could be dispensed either manually or with fixed piping and equipment; and would be usable, with modifications, for nonpolar flammable liq-

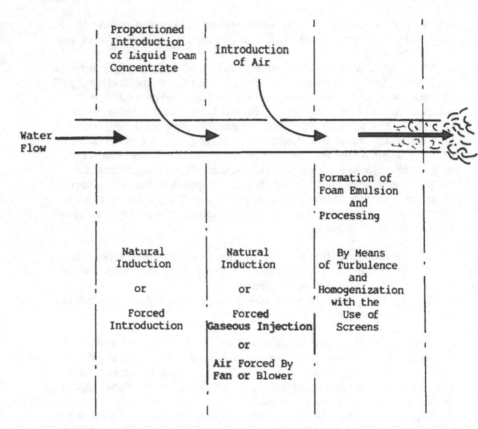

FIGURE 6.22 The three stages of fire-fighting foam generation.

uids, ordinary solid fuels, and polar flammable liquids. But, as
with water, electrical hazards exist but, admittedly, to a smaller
and smaller extent as foam expansion ratios exceed 500:1. There
is not now enough information to speak further.

Foams have been put to many uses, other than fire fighting,
such as dust suppression, crop treatment, froth flotation of
ores, treatment of fabrics, cleanup of radioactive spills, density control of drilling "muds" in oil and gas well drilling operations, and others. A scholarly article, "Aqueous foams" (*Sci.
Am.*, 1986, Vol. 254, No. 5), has been written by J. Aubert,

A. Kraynik, and P. Rand of the Sandia National Laboratories in Albuquerque, New Mexico, where continuing research on aqueous foams is being conducted.

Aqueous foams are being examined for their ability to decrease the pressure generated by the shock wave of an explosion. This dramatic ability arises from the destruction of the foam surrounding an impending explosion before its actuation. This unusual ability is due to the foam destruction surrounding the explosive: in particular, the shock wave energy is consumed in converting what was foam into minute drops of water. Additional energy is then absorbed by vaporization. All together, it has been determined that as much as 90% of the generated pressure is reduced and it makes itself evident in that particles propelled by the shock wave are virtually slowed down to low levels. This technique is used in coal mines by covering the coal face, that is being excavated, preparatory to detonation. The pressure of the explosion is reduced in the direction of the mine, and the coal dust is trapped, thus reducing the chance of a secondary explosion.

Data were collected from a test that was conducted using a 5-lb charge of C4 military explosive. The charge was covered with 800 ft^3 of 200:1 expansion ratio foam (from Table 6.7, 0.313 lb water/ft^3, therefore, 250 lb of water was in a state of foamy suspension). From Figure 6.16, we see that 1500 BTU/lb of water can be absorbed by evaporation and superheating to 1000°F (538°C). Hence, 375,000 BTU can be substantially absorbed in a very short period. It was during this latter test that it was determined that the severity of the explosion was reduced by about 90%. This capability of foam has attracted the worldwide attention of fire and police departments.

Modification to Reduce Surface Tension
(Water-Soluble Liquid Fuels)

If aqueous air foam is applied to a polar solvent (miscible in water), such as ketones, esters, alcohols, ethers, and organic acids, the foam will rapidly disappear because the liquid fuel will "wick" up through it, wetting it down to nothing. To provide for this problem and keep the foam in a stable condition the following conditions are necessary:

1. With protein-based foam concentrates, incorporate fatty acids and soluble metallic salts of such type that insoluble metallic soaps are formed when diluted into the main water stream.

These insoluble soaps protect the foam bubbles from being pen-
etrated by the polar solvent liquid fuels. There is, however,
a time limitation of about 0.5 to 1 min after mixing because of
the coalescing action of the soap curds, which would nullify the
effectiveness of the foam. This limitation prevents the pumping
of solutions from the proportioner for any substantial distance.

2. With aqueous film-forming foams (AFFF) a cohesive polymeric additive has been added (see Fig. 6.20). The additive
is a proprietary product and cannot be defined more specifically. In the type I situation, there is the condition, which was
illustrated in Figure 6.19, in which the AFFF—water film directly rested on the nonwater-miscible fuel surface. However, the
type 2 film must also remain on top, but it cannot directly contact the polar solvent because the film is dissolved. To prevent
this a cohesive polymeric additive is added whose properties include not being soluble in the polar solvent. The result is a
barrier film that is sandwiched between the AFFF—water film
and the polar solvent. Maintenance of this condition is assured
by surface tension phenomena. The films are always extremely
thin (about 0.013 mm) which prevents film dilution.

Diammonium Phosphate Solution

Diammonium phosphate additives have been used for fire-retardant treatment of paper, wood, textiles, and all vegetable matter,
in general. It is a white crystalline powder, extremely hygroscopic, that gradually loses 8% of its weight as free ammonia upon exposure to air, converting it to more stable monoammonium
phosphate, which is the main ingredient of multipurpose dry-
chemical extinguishers. Diammonium phosphate is extremely soluble in water to the extent that saturation is reached when it
is dissolved, by weight, in a ratio of 1:1.7 parts water at room
temperature. The mixture is very viscous, and the solution is
slightly alkaline.

As an indication of its effectiveness in treating ordinary materials to render them flame-retardant, a solution is prepared,
usually in large quantities that consist of 100 lb dissolved in 50
gal of water. This solution, when applied by spraying or dip-
ping, will be very effective if the original treated fabric becoming 10% heavier. Such a solution is usually applied at a rate of
1 qt/100 ft^2 of fabric area. Standard ignition test result in only the inevitable formation of char but with no flaming or after-
glowing of surrounding material.

Figure 6.23 illustrates the pattern of thermal decomposition of diammonium phosphate when applied (in solution form) to burning class A combustibles (see Fig. 6.8). After the release of ammonia gas the material is dehydrated to form a glassy, infusible clear coating that remains unchanged until at a bright red heat, about 1000°C, it volatilizes. The water itself exerts its own independent heat suppression but, upon vaporizing, leaves a residual fire-retardant coating and, as such, provides a lasting effect rather than a transient one. This residual coating is soluble and can be washed away after the fire.

Bentonite or Borax Slurry

Bentonite and borax slurries are being used very successfully in combating wild fires in brush and forested areas. Diammonium phosphate has also been used for this application but, for economic reasons, bentonite is the present favorite.

Bentonite is a colloidal hydrated aluminum silicate (clay) found in the United States midwest area and in Canada. It is a special clay, geologically known as *montmorillonite*, $Al_2O_3 \cdot 4 SiO_2 \cdot H_2O$. It has the property of forming highly viscous suspensions, or gels, with more than 10 times its weight of water. This gel-forming property is increased by the addition of small amounts of magnesium oxide. Although, as with other clays, it is not soluble in water, but it swells greatly by absorbing water to form gels. It has many uses, one of the most interesting of which is to make a drilling "mud" in oil and gas fields. In fighting forest fires, it is considered a short-term fire-retardant, and a slurry in water produces a heavy coat of water that is effective as long as the water is present. After the water evaporates, there is no further fire retardancy.

Borax slurries also have been used for the same purpose as bentonite. The resemblance is similar in that dehydration is the reason it is effective and also because no chemical action is involved in the development of fire retardancy. There is no gel-forming action as with bentonite. Sodium tetraborate, $Na_2B_4O_7 \cdot 10 H_2O$ is a natural product that is moderately soluble in water: approximately 10% as soluble as diammonium phosphate. Its action as a fire suppressor can be explained as follows:

1. The basic material has a high content of "water of crystallization," to the extent of 47% of its weight. When

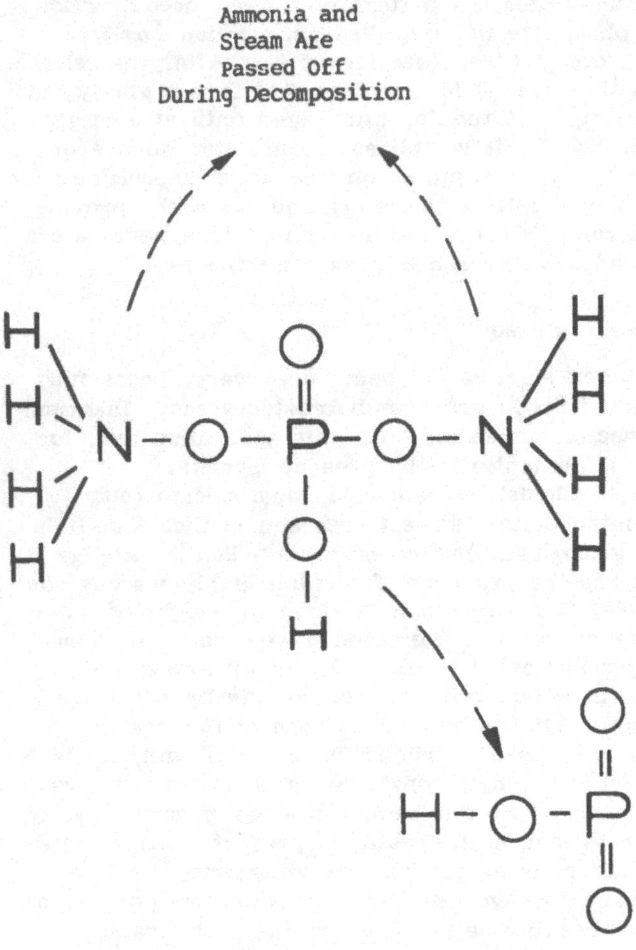

Ammonia and
Steam Are
Passed Off
During Decomposition

FIGURE 6.23 Thermal decomposition of diammonium phosphate.
It decomposes to meta-phosphoric acid at fire temperatures. It
is a glassy, infusable, clear coating that is unchanged up to
bright red heat. At higher temperatures, it volatizes to phosphorus pentoxide.

heated to 212°F (100°C) it begins to lose this water, and by the time the material reaches 576°F (320°C), the water of crystallization is totally removed. This action, together with the water content of the slurry, serves to supply the cooling action.

2. The way the sodium tetraborate crystals release the water of crystallization is interesting. After the main body of the water has evaporated at 212°F (100°C), the remaining crystals of the material release the relatively large amount of water of crystallization by "exfoliating" in a manner similar to the release of internal moisture in corn seeds when popcorn is made. The anhydrous sodium tetraborate will melt at 1364°F (740°C) and hence, a tenuous puffed film is formed. This covering also can be washed away. No permanent fire retardancy is developed.

Thickened Water

We have touched on the advantages of increasing the viscosity of water to counter the rapid runoff. A chemical means to accomplish this action is through the use of semigelatinous coatings over the burning solid.

There are a number of either naturally occurring or synthetically produced plant polysaccharide derivatives, normally produced by plants through photosynthesis from atmospheric carbon dioxide and water, resulting in the formation of cellulose and starch. There are an enormous number of different polysaccharides in nature. One of them, in particular, is of immediate interest: alginic acid, which is derived from seaweed. It is used as a thickening agent for foods and oils and in the manufacture of mucilage, as well as other agents.

A synthetic derivative has come into popular use for combating forest fires. It is derived from cellulose that has been treated with an alkali and then, subsequently, with sodium chloroacetate, producing a white granular powder, named sodium carboxylic methyl cellulose, that is nontoxic and soluble in both hot and cold water. It can be easily modified to produce the desired viscosity and is stable between the pH limits of 2 and 10. Concentrations are about 1% by weight. It has wide use as a thickening and suspending agent, in foods (especially ice cream), adhesives, and many other applications.

Although both of the aforementioned substances can be stored
in solution form for a limited time under pure conditions, it is
recommended that solutions be prepared in batch form preparatory to being used to prevent "peptizing" or thinning out.

Water Plus an Alkaline Salt

Saponification, the soap-making reaction resulting from the mixing of animal and vegetable oils with alkaline salts, has long
been known. A clever use of this reaction has been adopted to
extinguish a very particular type of fire; the burning of animal
and vegetable oils on the deep fat fryers, grills, broilers, and
associated cooking equipment such as used in restaurant kitch-
ens and galleys.

Animal fats and vegetable oils, specifically, are the "triglyc-
erides" of fatty acids. When reacted with an alkaline salt,
glycerin, steam, carbon dioxide, and soap result. It is interesting that when a water solution of an alkaline salt of, say,
potassium carbonate or potassium acetate is directed into the
burning oil at about 600°F (315°C), instead of a steam explosion
an immediate nonflammable fluid soap foam, containing steam,
carbon dioxide, and glycerin, forms that is stable and spreads
quickly over the burning surface.

There are other methods, namely, the application of alkaline
dry chemicals, that can also produce this same effect but not
as efficiently nor as cleanly, which is important in food-serving
establishments.

Halogenated Hydrocarbons

Earlier in this chapter, the gaseous halogenated hydrocarbons
(Halons) were described, particularly for their unique ability to
interrupt the chain reaction of combustion. These gaseous halogenated hydrocarbons are liquid only when they are stored under pressure within containing tanks and expand to a gaseous
state when released to the atmosphere.

There is only one current liquid halogenated hydrocarbon
(Halon 2402), which has a boiling point of 117°F (47°C) at atmospheric pressure, that is in use. If you turn back to Table
6.5, you will note that the first five listed Halons are all liquids, but only Halon 2402 is currently in use because it is both
less toxic and more stable than Halons 104, 1011, 1202, and
1001.

However, it must also be noted that all Halons release toxic decomposition products, such as carbonyl bromide and hydrogen fluoride, and exposure to them must be as short as possible. As with the gaseous Halons, Halon 2402 is also safe for use against live electrical apparatus.

Because it is a liquid in its normal state, Halon 2402 can be deployed in a manner similar to that of a water stream. Its low surface tension (25 dynes/cm) allows it to penetrate solid fuels such as bales, mattresses, upholstery, wood cribs, and the like. It vaporizes very quickly, because of its extreme volativity. The same extinguishing action exists as for Halons 1301 and 1211. The high liquid density of the halogenated hydrocarbons and their potency in fire extinguishment result in their occupying the minimum space for protection of a given hazard.

In summarizing the halogenated hydrocarbons, it becomes apparent that the choice of which agent to use depends upon the application. Halon 1301 is best utilized as a gaseous flooding agent within a confined space and Halon 2402 is best utilized as a local application, i.e., directly as from a portable or hose line extinguisher. Halon 1211 is in an intermediate position but has the most favorable economic aspect.

Synthetic Fluids

The term "synthetic fluids" applies to a very limited number of compounds that burn with great difficulty and emit relatively large amounts of carbon, both as a solid residue on the fuel and in the smoke. Because the carbon does not melt, it remains in situ until oxidized; however, if the agent is applied liberally enough, the high thermal conductivity of carbon serves to diffuse the heat over an expanding area and works toward the elimination of concentrated "hot spots."

These agents can be used with only some degree of effectiveness against combustible metals which usually occur in a state of division ranging from powder upward toward granules and shavings. The temperatures produced by burning metals are generally much higher than temperatures developed by hydrocarbon fuels. Burning metals belong to a genus of fuels different from the carbon-bearing fuels. Whereas the latter develop flame temperatures of about 3000°F (1700°C), metallic fires are virtually flameless; are of the glowing mode, as described in the first chapter; and burn at temperatures at least 2000°F (1100°C) higher. They present specific problems because they cannot be extinguished by the usual agents, such as water, carbon dioxide, steam, or nitrogen atmospheres, in which ordinary combustibles or flammable liquids would be incapable of burning.

The burning properties of metal fires cover a wide range. Burning titanium produces little smoke, whereas burning lithium smoke is very dense. Some moistened metallic powders, such as zirconium, can actually explode. Sodium, potassium, and sodium potassium alloy (NaK) flow while burning, whereas calcium does not. Some metals, as for instance uranium, acquire an increased tendency to burn after prolonged exposure to moist air. This subject is thoroughly covered in the NFPA *Fire Protection Handbook* in a manner too detailed to cover in this text. Ignition temperatures are subject to the state of subdivision but, generally, are approximately the same as the melting point. If the metals are mixed with mineral oils and combustion is initiated, the oil burns selectively until it is ultimately consumed and, subsequently, the metal burns. This fact has sometimes raised the prospect of inundating the burning metal with a heavy oil, causing a severe "secondary" fire, which if a sufficient rate of oil is applied, will enable its extinguishment by the usual conventional means of using water, carbon dioxide, or dry chemicals. This method is not without its hazards, however, and is mentioned only as a possible expedient in the absence of a better means. It is not considered an approved method.

In a similar vein, extinguishment has been accomplished using a 50:50 mixture of lubricating oil and either carbon tetrachloride (Halon 104) or chlorobromomethane (Halon 1011). No subsequent extinguishing methods are needed. The dense black smoke and the emission of toxic combustion products are considerable.

There are a few organic phosphates of high molecular weight, such as tricresyl (or tritolyl) phosphate, triphenyl phosphate, tributyl phosphate and dioctyl phthalate, that are very viscous or even semisolid materials and have to be rendered sufficiently fluid with organic solvents, such as alcohols or ketones, for dispensing and use upon metallic fires. Their use is known only in a general way, and no data exist to carry this thinking further. Their action after an initial burn-off is to form a heavy carbonaceous coating permeated through with the oxides of phosphorus which impart a full covering action. Carbon, because it is a good thermal conductor, serves to help dissipate the generated heat of the fire.

None of the aforementioned means have gained any recognition from approved agencies. It would appear that the only agents that are approved are solids (item 19, Table 6.4).

Alkaline Dry Chemicals

The past 40 years have seen the development and vastly increased use of dry-chemical extinguishing agents. Together with the Halon fire extinguishing agents, previously described, they comprise the only extinguishing agents that perform by breaking the combustion chain reactions. The rapidity with which this is accomplished has always been of intense interest because no cooling or oxygen dilution occur. Items 14, 15, 16, and 17 (see Table 6.4) are loosely related in that they are all alkaline. Their extinguishing mechanisms are similar enough to illustrate them, in the interest of simplicity, with Figure 6.24. For all four compounds, the cationic portion of the dry chemical molecule, sodium or potassium, is fractured off by thermal decomposition and reacts sequentially with the H^* and OH^* free radicals (chain carriers) developed in the combustion process, culminating in the formation of steam and the reemergence of the original cationic fragment as an active free radical, thus continuing the chain-breaking action. Also, with all four, the anionic portion of the dry chemical molecule is fractured off by thermal decomposition and reacts with the H^* free radical to form stable end products, thereby removing it from the field of combustion.

To understand the relative status of the various dry chemicals, it is necessary to view them in an historical context. During the early period of the petroleum industry, there was little need for the gaseous and lighter fractions of petroleum. With the advent of the internal combustion engine and its dependence upon gasoline, the need for handy and reliable fire extinguishers became imperative. Chemical foam and carbon tetrachloride became a means of control but difficulties existed because of mechanical and maintenance problems. It also became known that certain dry solid materials such as borax and sodium bicarbonate appeared to be effective, but practical difficulties prevented any general use. Meanwhile carbon dioxide came into general use after World War I.

Sodium Bicarbonate-Based Dry Chemicals

The use of borax developed from the knowledge that it was the means of coating burning ordinary combustibles with boric acid

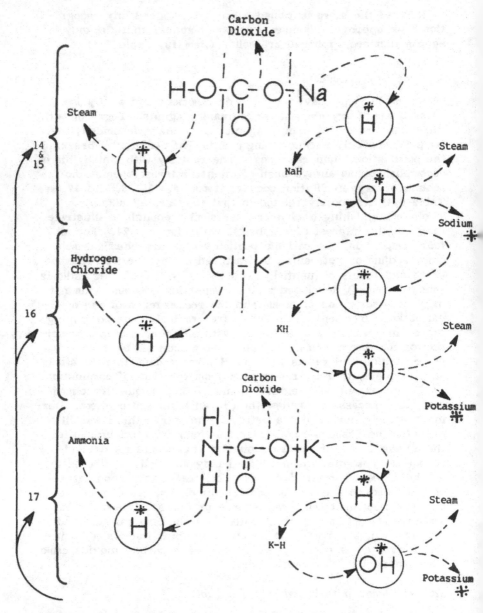

FIGURE 6.24 Extinguishing mechanisms of the alkaline dry chemicals (refer to items 14–17, Table 6.4).

and its anhydride, boric oxide, which are known fire retardants.
However, sodium bicarbonate was discovered to have the property of actually extinguishing flames. This latter feature was
superior for two reasons: the first was its nonhydrated form,
whereas borax was highly hydrated and lumpy, the second was
for economic reasons because sodium bicarbonate is relatively
inexpensive, a result of its wide use in the food industry.

However, the processing of sodium bicarbonate to impart superior moisture resistance presented a serious problem: when
stored within a closed container and exposed to warm environments at temperatures in excess of 70°F (21°C) the following
reaction occurs increasing as the temperature rises:

$$2 \, NaHCO_3 \rightarrow Na_2CO_3 + H_2O + CO_2$$

Hence, both moisture and carbon dioxide are emitted. Whatever moisture had been initially imparted became overcome by the
moisture emission from within the molecules causing cohesive-
ness, lumping, and unsatisfactory operation of the extinguisher.
These problems were partly resolved by the use of superior
grades of insoluble soaps, such as zinc and magnesium stearate.
However, moisture still posed a problem that had not been fully
solved owing to the never-ending hydrolysis reaction between
the stearates and the sodium bicarbonate.

*Potassium Bicarbonate- and Potassium
Chloride-Based Dry Chemicals*

In 1952, I became involved in a most crucial project, namely,
the use of silicone resins to treat sodium bicarbonate to provide
a total moisture repellancy, despite any evolving moisture, to
all of the individual granules, thereby preventing the former
troublesome adhesion problems. The process was set forth in
the following patents:

2,866,760	United States
607,371	Canada
824,107	United Kingdom

and was entitled "Compositions and Methods for Treating Par-
ticulate Solids." This process made it possible to develop newer dry chemicals, such as potassium bicarbonate-, potassium
chloride-, and potassium carbonate-based types, which are all

notably more hygroscopic than sodium bicarbonate. Even the
monoammonium phosphate-based dry chemical, which is abnormal-
ly hygroscopic, was successfully treated. Dry chemicals had finally come of age. The potassium-based dry chemicals are somewhat more effective than the sodium-based dry chemicals because potassium has a lower ionization potential. The potassium
bicarbonate and the potassium chloride dry chemicals are equal-
ly effective; however, the latter agent gives rise to hydrogen
chloride emission that, in the presence of steam, is corrosive.
However, the chloride is a naturally occurring material and,
hence, there are economic reasons for using it instead of the
bicarbonate, which requires chemical processing.

Potassium Carbamate-Based Dry Chemical

The potassium carbamate-based dry chemical is the latest and
most potent of this group for fire fighting. It was developed
in Great Britain and is the reaction product of molten urea and
potassium bicarbonate that, after being cooled and ground down
to size, is treated the same way as the other dry chemicals.
This dry chemical is the most potent of all the various types,
as gauged by the Underwriters' Laboratory testing described in
Chapter 1, and requires use of only about one-half the amount
of the other potassium-based agents on given test fires.

All of the dry chemicals described so far have one common
failing: namely, their inability to extinguish class A fires involving ordinary combustibles. As Table 6.4 shows, they do
not vary in fire classification, rather, they vary only in degree.

Monoammonium Phosphate-Based Dry
Chemical (Multipurpose)

One of the most important developments relative to the dry
chemicals occurred in Germany. Attempts were made to impart
a characteristic to obtain a coating or covering of the solid fuel
with a high fire resistance value in addition to the pronounced
ability of dry chemicals in extinguishing flames. One of these
efforts was directed toward a coating that would "intumesce,"
as certain fire-retardant paints do, on the burning surface.

The most successful of these various efforts involved the use
of monoammonium phosphate, which provides an effective covering on the burning solid while simultaneously directly combating
the flames. It does not provide any cooling effect, but it has
found wide utility as small, light-weight portable extinguishers

in place of bulky, pressurized water extinguishers and the out-
dated soda/acid extinguishers. By referring to Table 6.4, it
will be seen that the monoammonium phosphate-based agent has
capabilities beyond those of the other agents.

Figure 6.25 illustrates diagrammatically the manner in which
monoammonium phosphate decomposes within a fire. The ammonium radical (NH_4) is fractured from the basic molecule and
combines with a hydroxyl free radical (OH^*) to form free ammonia and steam. The remaining portion of the original molecule
combines with a hydrogen free radical (H^*) to form orthophos-
phoric acid that, upon dehydration, forms steam and metaphosphoric acid, which takes the form of a tough, thin fire-resistant
coating on all surfaces. Metaphosphoric acid was described earlier in this chapter and in Figure 6.23.

Finally, *never* mix this acidic dry chemical with any of the
alkaline dry chemicals. They mutually deteriorate and release
carbon dioxide.

Granular Graphite or Salt

Seventy-five percent of all the 92 natural chemical elements are
metals as defined by chemists: they are electropositive and
form a base or alkali when reacted with the electronegative hydroxyl (OH^*) free radical. All of them can burn under certain
conditions, and most are very difficult to ignite. In their so-
called solid state they usually can be expected to become vulnerable to ignition near their melting point. The combustibility
of all metals appears to increase as the average particle size is
reduced, but other variables, such as moisture content, also
affect the ease of ignition. One extreme example of this behavior can be shown by zirconium, which in its solid massive form
can withstand a temperature of approximately 2500°F (1370°C)
before ignition, whereas clouds of dust in which the average
particle size is 3 µm (0.025 in.) can be ignited at room temperature. Instances of spontaneous combustion have even been
reported. Combustion of zirconium dust in air is stimulated by
the presence of limited amounts of moisture, a maximum of from
5 to 10%. The metals hafnium, tantalum, thorium, uranium,
and plutonium (not a natural metal), all appear to have similar
properties. Titanium differs in that the dust, both dry and
damp, will ignite spontaneously when temperatures exceed.630°F
(332°C). Furthermore, titanium dust can be ignited in carbon
dioxide and nitrogen.

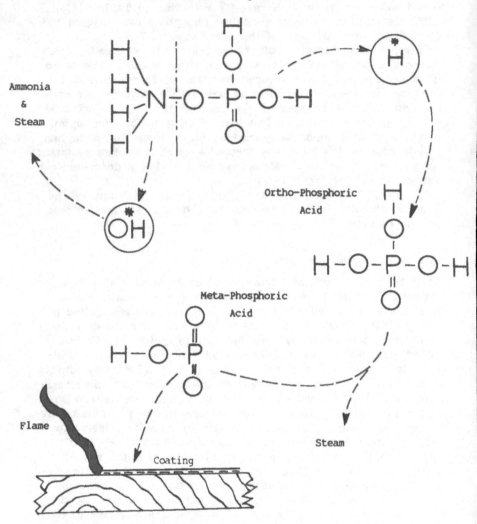

FIGURE 6.25 Extinguishing mechanism of multipurpose monoammonium phosphate dry chemical (acidic) (refer to item 18, Table 6.4).

Of the 92 natural elements, 69 are metals, 10 are nonmetals, 2 are liquid, and the remaining 11 are gases. The term, nonmetals, applies to those solids that are electronegative and form acidic materials when reacted with the hydrogen free radical (H^*). All of the 10 nonmetals are combustible, and their ease of combustion is enhanced by the reduction in particle size, all the way down to dusts. Furthermore, they are more easily ignited in their solid or massive state compared with the metals. The nonmetals principally comprise the elements carbon (which we have previously covered in this text), phosphorus, and sulfur. Other nonmetals may be occasionally encountered including boron, silicon, arsenic, selenium, and iodine.

The interested reader is referred to the NFPA *Fire Protection Handbook* for pursuing this matter in greater depth as well as for obtaining a large bibliography.

Table 6.4 lists the use of granular graphitized coke-based or sodium chloride-based (item 19) chemicals as being suitable for use in extinguishing metallic fires. As broad-based as the types of metallic fires appear, these are two agents that have wide utility.

First, the granular graphitized cube is treated with an organic phosphate additive that, although it burns with difficulty, forms copious amounts of free carbon, which serve to seal the interstices between the granules and prevent oxygen from reaching the covered burning metal. The agent is not capable of being dispensed by a pressurized extinguisher. It must be applied manually by either a shovel or from cardboard tubes in adequate amounts as observed by the operator. Second, it cannot adhere to inclined or vertical surfaces. Third, and perhaps most important, is that carbon has a very high level of heat conductivity and, thereby, serves to dissipate the generated heat. This agent is usually referred to as G1 (Pyrene) or Metal Guard, the patents for which have expired.

The second agent is referred to as "salt" in Table 6.4. It is finely pulverized sodium chloride that has been rendered moisture repellant and then blended with a small amount of tricalcium phosphate to assist its flowing property. A thermoplastic material, such as vinyl chloride, is added to bond together the sodium chloride particles into a sticky semisolid mass upon the burning metal. This agent, identified in the trade as "Metyl-X," can be gently discharged from a pressurized extinguisher so that burning metal particles are not scattered. The agent does adhere to inclined and vertical surfaces and is electrically nonconductive. However, it does not conduct heat away as does the granular coke.

These two agents can be successfully applied to the eight metals, as shown in the upper part of Table 6.8. The first four metals, namely, the alkali metals, such as lithium, sodium, potassium, and the liquid alloy of sodium and potassium known as NaK, all are characterized as being highly reactive with water, releasing free hydrogen and oxygen which reignite in an explosive manner. Such fires should *never* be attacked with water, and the dry chemicals, carbon dioxide, and Halons are totally ineffective. These metals are highly reactive with the air even when in a dry state. They are safely stored submerged in kerosene. They are further characterized by their low melting points and, hence, are in a liquid state when burning. Note that each of these metal fires are, therefore, referred to as "spill" fires, which generally have been considered to be from 1/4- to 3/8-in. deep. Greater depths will not only make fire fighting doubtful but probably unsuccessful. Ignition temperatures are all low enough to be considered pyrophoric when moist. The alloys, referred to as NaK, contain various percentages of sodium and potassium that vary in melting points from 13°F (−11°C) to 66°F (19°C). These alloys are employed as heat-transfer liquids in certain kinds of nuclear reactors because of their exceedingly high unit heat capacity. A leaky pipe or tubing fitting that is conducting NaK will result in an immediate spontaneous fire.

The remaining four metals, aluminum, magnesium, titanium, and zirconium, are listed in the order of increasing ease of ignition, as the table will show. In powdered form titanium and zirconium are pyrophoric, particularly when slightly moist. Extinguishment can, under special conditions, be successfully accomplished but not without a pyrotechnic display that is hazardous to fire fighters. Relatively large water discharge rates must be employed with use of straight-stream nozzles and at ranges in excess of 30 ft. All fire personnel involved must wear full protective clothing and face masks. Dry chemicals and carbon dioxide are ineffective. Halon extinguishing agents will actually cause explosions because of the reactivity between the released halogen and the burning metal.

The lower part of Table 6.8 is concerned with the solid combustible nonmetals. This separation is to distinguish them from the previously discussed solid combustible metals. The periodic table of elements contains 10 solid nonmetals, the most common

TABLE 6.8 Comparative Combustibility and Extinguishing Agent Compatibility for Various Metals and Nonmetals

	Ignition factors	Extinguishment	Item 19 (Table 6.4)
Some common metals and their hydrides			
Li[a]	Lithium spill ignition MP 367°F (186°C)	NEVER use water	
Na[a]	Sodium spill ignition MP 208°F (98°C)	Explosive reaction	
K[a]	Potassium spill ignition MP 144°F (62°C)	Dry chemical, CO_2, and Halons not effectual	
NaK[a]	NaK spill ignition MP 13°F (−11°C) MP 66°F (19°C)		
Al	Particularly as a powder or turnings; ignition approx. 1200°F (650°C)	Use water ONLY in copious quantity; use straight-stream nozzles in excess of 30-ft range; use full protection; dry chemicals, CO_2, and Halons cause explosions	
Mg	More hazardous than aluminum; ignition approx. 960°F (510°C)		
Ti[a]	Titanium powder or turnings; large masses also susceptible		
Zr[a]			
Some common nonmetals			
C	Carbon, ignition in excess of 806°F (420°C)	Use water in any of its forms, Halons or ABC dry chemical	
P	Phosphorus, ignition (white)[a] 104°F (40°C) (red) 500°F (260°C)	Always use water spray at close range	
S	Sulfur, ignition at MP 235°F (112°C)		

The Li, Na, K, NaK items are bracketed together as "Alkali metals in general."

[a]Pyrophoric, particularly when slightly moist.
MP = melting point temperature.

of which are carbon, phosphorus, and sulfur. Boron, silicon, arsenic, iodine, and the like are seldom encountered in their elemental state.

The most common, carbon, has been covered. The element phosphorus has two allotropic forms: the white form is pyrophoric and extremely poisonous, the red form is not pyrophoric and has an ignition point of about 500°F (260°C) but, if heated and then cooled, reverts to the white form. The element sulfur ignites at its boiling point of 235°F (112°C). Both burning phosphorus and sulfur emit highly toxic and choking fumes.

In all of these three instances of nonmetals, water provides the preferred means of extinguishment. The previous agents, "G1," Metal Guard, and Metyl-X are ineffective on phosphorus and sulfur fires.

As can be deduced from the foregoing discussion, the fire behavior of the eight common combustible metals (and their hydrides) and the three common combustible nonmetals are highly individualized and call for a degree of sophistication in mounting an attack. These materials exist primarily in industrial or military installations that maintain their own special fire brigades. The hazards they present are many and different and require expert training.

Many nonapproved and improvised materials have been used such as coal, sand, dirt, foundry flux, soda ash, and so on, with varying success. However, the material used must always be dry to protect the firemen.

SUMMARY

As one reviews the 19 separate classes of extinguishing agents in Table 6.4, it becomes most obvious that each one has its advantages and disadvantages. It must also become obvious that when two agents are used in consonance with each other, each having properties that compensate for the deficiencies of the other, we have a synergistic situation. When this procedure is followed, extinguishment is performed more efficiently with far less of each agent used than if they had been used separately. Various examples can be cited, namely, the concomitant use of dry-chemical agents with detergent—water solutions or the use of certain halogenated hydrocarbons with detergent and AFFF solutions. In particular, the joint use of "Purple K" potassium bicarbonate-based dry chemical (item 15, Table 6.4) with AFFF (item 6 or 7, Table 6.4) has become standard procedure for many naval vessels and airport operations, both military and civilian. Also, some notable performance has been observed with the joint use of multipurpose monoammonium phosphate dry chemical with detergent—water solutions.

The total objective has been to obtain as complete as possible coverage of the A, B, and C fire classifications, as well as of the four basic fire extinguishing mechanisms. Each of the two agents, therefore, complement each other.

However, metallic fires remain excluded from these binary combinations and require their own specific treatment.

The foregoing summary very definitely points to the future trends of fire fighting. More efficient extinguishment, shorter periods of required extinguishment, faster means of accomplishing rescue of trapped individuals, lessened damage to the environment, all will result. The fire-fighting potential of each and every member of the fire combat force will be increased. The increased performance of lighter-weight, mobile apparatus will help to shorten response time. The improved utilization of human and material resources are essential in protecting our every-increasing fire hazards.

7
Environmental Considerations

So far we have viewed the fire phenomenon as a visible, readily recognizable, chemical reaction, having either a glowing or a flaming character dependent upon the fuel and the surrounding environment. Therefore, decisions could be made to determine the preferred type of extinguishing agent(s) to use as well as the manipulative skills required.

The interplay between the immediate environment and the area of combustion is of considerable importance and will be covered in this text from three standpoints.

1. The development of hidden combustion and its hazards.
2. The thermal effects of flaming combustion upon the surrounding environment.
3. The toxic effects of the gases and smoke in the surrounding environment.

SPONTANEOUS IGNITION

The term "hidden combustion" is most often referred to correctly as spontaneous combustion, defined in *Webster's Dictionary* as "ignition in a thermally isolated substance, as oily rags or hay, etc., caused by a localized heat-increasing reaction between the oxidant and the fuel." It is in this sense that the term is used in this text to distinguish it from "autogeneous combustion," a term that implies that the fuel burns in a manner that is self-produced or independent. The etymology of these terms may seem obscure because both of them refer to ignition that is reached by no external means, such as pilot flames or sparks. Nevertheless, the two terms have been used interchangeably, and so we must define them more clearly. The term "autogenous" can best be associated with the following examples:

1. Autoignition temperature (AIT) as described in Chapter 2 and illustrated in Figure 2.6.
2. Certain examples, such as lithium, sodium, potassium, rubidium, and cesium (metals of the Group 1 Periodic Series), titanium and zirconium powders (particularly if slightly moist), and the nonmetal, white phosphorus. All of these elements ignite in open air because of their high order of mutual chemical activity.
3. Reactions between two separate chemicals which, by their relative chemical activity, erupt into flames and perhaps even explode when brought into contact with each other. There are many examples.
4. Shock-induced combustion by which chemical instability can cause explosions.

THE BASIC MECHANISM OF SPONTANEOUS COMBUSTION

In contrast, spontaneous combustion, as we use the term, refers to an insidious, slowly expanding combustion process hidden from view. Six classic examples are illustrated in Figure 7.1, each of which will be described in some detail. The fuels that are the basic cause for this phenomenon are all solid organic substances that react directly with air, most often promoted by the presence of moisture. The reactions of this type are in the "solid-to-gas" category that begin at normal temperatures within a sheltered area and slowly keep developing to higher temperature levels through oxidation. The NFPA *Fire Protection Handbook* contains a long listing of the organic materials subject to spontaneous combustion that includes such disparate materials as charcoal, bituminous coal, cattle feed,

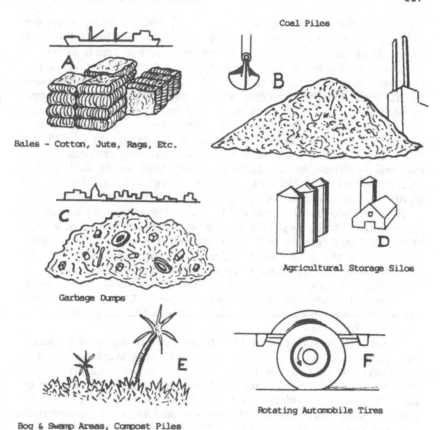

Coal Piles

Bales - Cotton, Jute, Rags, Etc.

Garbage Dumps

Agricultural Storage Silos

Rotating Automobile Tires

Bog & Swamp Areas, Compost Piles

FIGURE 7.1 Six classic examples of spontaneous combustion.

fertilizers, hay, cocoa beans, manure, oil-soaked rags, meals derived from alfalfa and corn, and so on. The origin of the heating process can be attributed to various causes, such as the
proximity of heated water, steam pipes and chemical fertilizers.
Another source of heat is the oxidation that is frequently caused
by bacterial biothermogenesis and is instrumental in the decay,
fermentation, and rotting of vegetable matter. These processes
are promoted by moisture and are stimulated by a rise in temperature. Bacterial action ceases when temperatures of 160 to
170°F (74 to 77°C) are reached. If we consider that many
chemical reactions double their rate for about every 20°F (11°C)
rise in temperature, the bacterial factor can account for an
eight- to ten-fold increase in the rate of oxidative reactivity.

Another source of initial oxidation is the attack by oxygen at
the reactive allylic positions in the molecules of unsaturated vegetable oils and fats as well as on discarded animal or vegetable
matter, with the consequent evil-smelling development of rancid-
ity. Linseed and tung oil have a special importance because of
their high content of glycerides derived from acids that contain
a number of double carbon (=C=C=) bonds. This process is not
too well understood, but there is no doubt that it presents serious fire hazards.

The evidence seems to be that these various forms of preheat-
ing become a prelude to a continuing and accelerating rate of
oxidation. When exposed to open air, all organic material,
whether liquid or solid, reacts with oxygen in an exothermic manner. The rate at which this proceeds is subject to many variables such as temperature, the nature and state of subdividision
of the substance, the presence of free moisture, the presence
of certain bacteria, and the degree of ventilation. All of these
factors defy quantification and, thus, any description of the
overall process must be expressed in general terms. Whether
or not this oxidation produces any discernible or recognizable
results, in spontaneous combustion, depends upon the shelter
offered within interior areas as well as the porosity of the combustible substance itself in permitting air to be present. For
example, a solid piece of wood cannot experience this sort of
combustion, a sizable pile of materials, such as sawdust, grains,
coal, and the like, may develop recognizable combustion in a
period of months, and a painter's rags containing linseed oil or
turpentine within a closed space will ignite spontaneously, in its
entirety, in 1 or 2 days. The governing principle in all these
examples, is the relationship of the rate of generated heat to the

rate of heat dissipation to the surrounding material. Spontaneous combustion can be defined as starting when the oxidating process, which began at normal temperatures, reaches a self-sustaining stage. Up to that point, endothermic decomposition has prevailed. However, at some point, the endothermic decomposition begins to diminish, and because of the higher temperature, exothermic decomposition becomes apparent. Table 7.1 presents a scenario of this complex matter in a far better way than can be accomplished with prose. This table illustrates the generalized sequence of the various steps in the oxidation of a cellulosic substance, which would be an example of most vegetable matter.

Figure 7.2 emphasizes a most important concept, that is, for spontaneous combustion to proceed it must be submerged within the combustible porous material at such a depth (critical depth) that, first, air can reach it and, second, that enough heat insulation is provided by the material overlay that the generated heat caused by oxidation is retained. The deeper the zone, the greater will be the heat retention *but* at a sacrifice in air supply. Conversely, the shallower the zone of overlay, the more readily air will penetrate *but* at a sacrifice in heat retention. The critical depth defines that region, measured down from the surface in contact with the open air, at which both conditions for continuing combustion are met. The larger the burning region becomes, the nearer it approaches the air boundary, at which point open flames will erupt.

Charcoal bears special mention. In its manufacture, because of its extreme porosity, great care must be exercised to keep it as dry and as cool as possible, as well as to prevent inadvertent contamination with combustible foreign material or oxidizing agents. If accidently wetted, it must be quickly removed because it is then more susceptible to self-heating than when dry. However, when dry it becomes very dusty, thereby presenting explosion hazards. When one refers to Table 7.1, it will be very obvious that as temperatures mount, carbonization increases and charcoal becomes a prominent result of the increasing tempo of oxidation which, together with the water vapor and combustion by-products, makes charcoal a major constituent of the fire.

TABLE 7.1 Approximate Progress of the Thermal Degradation of Cellulosic Substances

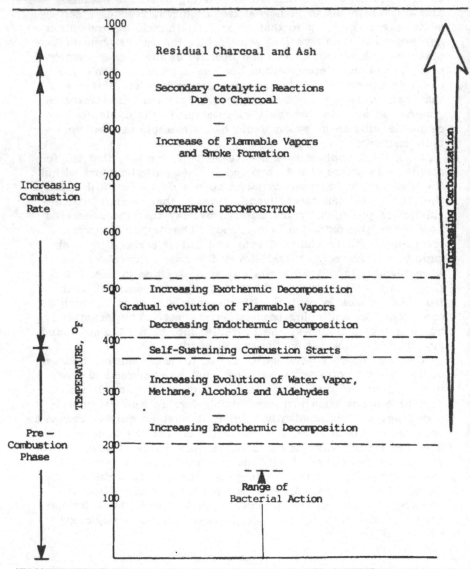

Fibrous, Granular

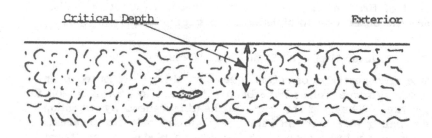

Critical Depth   Exterior

Expanding Combustion

Exterior

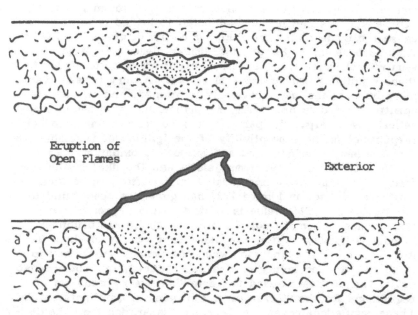

Eruption of
Open Flames   Exterior

FIGURE 7.2 Spontaneous combustion syndrome.

EXAMPLES OF SPONTANEOUS COMBUSTION

The six classic examples shown in Figure 7.1 will be viewed separately because each has its own characteristics in the development of fires originating from spontaneous combustion. They
are discussed now in alphabetical order.

Cotton, Jute, Hemp, or Sisal Bales (Case A)

The transportation of cotton bales has always presented problems. Various causes have been attributed to be the reason for
ignition. They include frictional heating caused by vibration of
the baling straps within improperly stowed bales in an area,
such as a railway boxcar, or perhaps to pieces of hot "tramp"
iron that inadvertently have been baled within the cotton. The
time interval of transportation in either highway or railway
transport is relatively short, a matter of a few days at most.
Yet, such instances have been reported but, in the main, the
experience is of a low order. Of more importance is the experience of many fires at sea when cotton was stowed in poorly
ventilated holds and usually subjected to heated conditions because the trade routes for shipping cotton originate from coun-
tries such as India, Egypt, and similar places of a tropical or
semitropical location. The time of exposure of cotton bales to
these conditions can last several weeks. A very possible cause,
in addition to the ones previously mentioned, and one that is
admittedly hard to prove but, nevertheless, has a sound theoretical bases (e.g., the painter's rag containing linseed oil or
turpentine) is the susceptibility of the ignition of vegetable oils
that are largely unsaturated and subject to oxidation.

Bales of cotton, jute, hemp, sisal, and the like, in the raw
form, do contain these unsaturated oils. Often, the critical
depth, as defined in Figure 7.2, has generally been found to
be about 6 in. This value is subject to the degree of compaction obtained when the cotton, jute, or such, was originally
baled. Originally, cotton was baled to a density of about 35
lb/ft^3, but it was found that when the density was raised to
50 lb/ft^3, the occurrence of fire was reduced (but not eliminated). The US Coast Guard has established a rigorous set of
regulations for the storage of cotton bales in merchant vessels.
These regulations cover the degree of separation from the deck,
the bulkheads, and between groupings of bales. The problem
has been recognized for a considerable time.

An interesting, but extremely unorthodox, method of extinguishing a smoking (but not flaming) bale of cotton, is to inject kerosene by means of a lance, internally "wetting" the cotton with the resulting vaporization serving to forcibly expel the air trapped within the bale. Smoking bales when thrown overboard alongside the vessel, and retrieved about 2 weeks later, have been known to start smoking again and actually flare up.

The only way for water to enter the bale from the exterior is to reduce its surface tension to overcome the water-repellant nature of the natural raw oils and resins within the raw cotton. It is only necessary that the wet water penetrate from 6 in. to some safe average. Whatever the cause of ignition, fire extinguishment can be promptly accomplished by opening the bale and extinguishing the hidden fire.

Coal Piles (Case B)

The storage of coal in piles, silos, or bunkers, and its influence upon spontaneous combustion has always been a subject of considerable controversy. This is because there are, at least, three separate and distinct chemically related causes and, at least, three more separate physical reasons, all six of which can combine to produce many different effects. They are as follows:

1. The higher the equilibrium moisture content of the coal, inherent within the coal itself, the more readily it is subject to combustion.
2. The higher the sulfur content of the coal, the more susceptible it is to combustion.
3. The higher the percentage of oxygen contained in a combined manner within the coal, the more susceptible it is to combustion. For example, the following is a listing of the susceptibility of different types of coal:

   Anthracite: no hazard
   Semibituminous: very low hazard
   Bituminous: intermediate
   Subbituminous: intermediate
   Lignite: most hazardous

4. Freshly mined coal is always the most vulnerable to oxidation.

5. Of particular importance is the matter of particle size,
    especially if the coal has a wide disparity of size, i.e.,
    large lumps mixed with finer ones, which is the usual
    form in which it is transported and stored.
6. The finer-particle size always presents more area to the
    air for oxidation, but if it is mixed in with large lumps,
    airflow is promoted through the pile and the matter of
    critical depth (see Fig. 7.2) comes to the fore. The
    variations in critical depth range widely from 1-1/2 ft to
    3 ft or more.

As for cotton bales, compaction of the coal pile is a very effective technique in retarding combustion. Also, if the pile is
scraped by a bulldozer, as well as compacted, another benefit
is derived by virtually keeping the pile from being just too long
in a static state. This latter condition can possible exist when
with coal supplies are temporarily threatened, as by strikes,
and extra-large supplies are purchased for a possible reserve.
In modern coal-fired, power-generating stations the coal is
moved by a multiple conveyor system to feed a multiple number
of silos, each of which feeds a coal pulverizer that by means of
primary airflow keeps the ground coal in a state of mobile suspension to the burners in the boiler furnace. A modern coal-
fired, steam-powered electric power station of 1500 Mw capacity
can be expected to require approximately 300 lb of coal per second, or 540 ton/hr, when operating at full capacity. The actual tonnage will vary with the quality of the coal. The individual coal silos will hold from 250 to 300 tons each. The coal
is transferred from the exterior pile to the silos in a manner
such that the residence time of the coal in any silo is minimized.
The entire action is truly a traffic-management problem. I was,
at one time, engaged in a particular investigation that will out-
line the overall problem. First, the coal was subbituminous,
containing from 6 to 7% total sulfur content. It was mined in
central Ohio and transported by automated rail transport directly to the power plant. The coal was then piled in an orderly
manner, preparing it for transfer to the interior of the power
station. On cool, damp days the coal piles would steam and
emit sulfurous vapors. In the winter, snow never remained upon the piles, whereas all of the surrounding terrain was covered. The transit time of the coal descending through the silos was to be limited to not more than 24 hr, although actually
the time was considerably less. Mechanical difficulties occasion-

ally interfered with the coal pulverizers, and a silo became in-
active. This difficulty became most serious, for spontaneous
combustion would start within the silo itself. This required
shutting down, emptying the silo into a pit, and thus causing
great problems. Steps were taken to counter this condition by
passing carbon dioxide or nitrogen from a refrigerated liquid
storage supply into the base of the silo and, thereby, inertly
inactivating the coal, because the gases passed away from the
top of the silo. Admittedly, the instance cited is extreme, but
it presented some very complex problems, which had a powerful
effect upon the environment. If we take the example of 300 lb
of coal per second, containing 6% combined sulfur, we come to
the horrendous answer of 18 lb of sulfur burning every second
and being converted to 36 lb of sulfur dioxide, which is the
root cause of "acid rain" that has become such a steamy political matter. Sulfur has become the villain in both spontaneous
combustion and atmospheric pollution.

Garbage Dumps (Case C)

Along with the world's expanding population and rapidly evolving industrial activity, our civilization is most certainly pollut-
ing its nest. The mounting sewage stowage and disposal problems have given rise to industries that specialize in this service.

The refuse and rubbish make up enlarging garbage dumps
that consist of an enormous potpourri of food wastes such as
rotting vegetable and animal matter of all types, metal, wood,
paper, cardboard, plastic, rubber products, leaves, plant and
grass clippings, paint, paint cans, aerosol cans, ad infinitum.
These garbage piles are exposed to the elements and become
subject to spontaneous combustion, and atmospheric pollution
becomes a vital matter of concern.

Continued burning creates a most hazardous problem in that
as the subsurface burning occurs it leaves behind a loosely
skeletonized residue of ash and noncombustible materials that
have no resistance to penetration. Hence, if a fire fighter
steps onto such a "booby trap" he will fall into a pit, causing
injury and burns. Discarded aerosol containers are likely to
either explode or "rocket." I know of a number of communities
in the greater New York metropolitan area where the garbage
disposal methods require that their fire service personnel and
equipment are in a constant state of readiness to attack garbage-
induced fires, in addition to being ready for emergencies.

The cause of these fires is the same as that of Case A: oxidation produced as a result of bacterial activity when living off of the nutrient content of the putrifying vegetable and animal matter.

As in all of these examples, the use of water, with the hope of achieving extinguishment, is both futule, time consuming, and wasteful.

There are two approved methods for disposal of garbage: (1) incineration under controlled conditions that minimize smoke and odor and keep all dust confined, and (2) land-fill operation where garbage is mixed with earth and, subsequently, compacted with bulldozers in a wide trench, in a layered fashion and, finally, covered with at least 2 ft of compacted earth.

Disposal by dumping at sea is not only unsatisfactory but also unsafe because of pollution. Open burning, done deliberately, is no longer legal in most states and is rapidly being forced out of use. Disposal by mere simple burial is also unsatisfactory because of the possible leaching of drainage into water resources. Pesticides that were buried in the 1930s have caused arsenic poisoning of well water some 40 years later.

Agricultural Storage Silos and Barns (Case D)

For years, it has been known that agricultural storage silos and hay bale storage in barns have been subject to the same bacterial degradation, with the help of moisture, as garbage. This interaction generates heat by oxidative means through bacterial action until the temperature reaches 160 to 170°F (74 to 77°C). As discussed earlier in this chapter, chemical reactions approximately double their rate for every 20°F (11°C) rise in temperature; thus, the bacterial factor accounts for an eight- to tenfold increase in the rate of oxidative reactivity and acts as a prelude to continued oxidation toward ultimate ignition. The air that is necessary to produce this oxygen is contained within the interstices of the granular or fibrous material as well as being supplied by natural ventilation.

Experience has indicated that spontaneous heating that can be detected becomes apparent during a period of 2 to 6 weeks after storage. The wide variation is due to seasonal effects of atmospheric moisture and temperature. The farming areas have experienced fire loss of harvested crops primarily because of the improper use of large quantities of water. As we have previously noted, water will only extinguish a class A fire when it actually reaches the seat of the burning fuel. Because a silo is essentially closed, except for top hatches and bottom unloading areas, there is no choice but to inject water through an open roof hatch, which itself is hazardous once flammable decomposition gases collect at the top of a silo. Water, when used in this manner, is ineffective for fire extinguishment, but it is highly effective in permanently damaging the silo, rendering the contents so sodden as to defy cleaning out. The danger also exists of the "water—gas" reaction, wherein moisture and incandescent carbon unite to form carbon monoxide and hydrogen gases as was detailed in Chapter 5, with particular emphasis as shown in Table 5.1, and Figures 5.2 and 5.3. There appear to be two general methods that have been successfully used in combating silo fires.

1. Use "dry ice," carbon dioxide in solid, broken up form, which has a temperature of −110°F (−79°C), and inject it into the top of the silo (providing the roof hatch can be safely opened). The dry ice will release gaseous carbon dioxide, which is 58% heavier than air and disperses through the silage, thereby displacing oxygen and allowing the silage to cool.
2. Use liquid nitrogen, which has a boiling point of −320°F (−196°C) at atmospheric pressure and expands into a gas that is slightly lighter than air. It should be injected with a "lance" at a low level (obviously below the fire).

Although it is virtually impossible to quantify the amount of either of these gases, we can only relate that for a 20-ft diameter x 50-ft high silo it can be roughly estimated that 600 lb of dry ice introduced from the top of the silo has approximately the same effect as 400 lb of liquid nitrogen. For silos that are either larger or smaller in volume, the amount of dry ice or nitrogen can be prorated directly.

Although this method of fire control has proved to be successful, we are still faced with cause of the fire, namely the unburnt dry silage. The proper procedure to be followed subsequent to the fire control is to empty the contents of the silo as rapidly as possible and move the contents to a safe place removed from surrounding combustibles.

With hay bales, the precaution is to not place them in large unmanageable piles but, rather, to set small piles apart from

each other, to enable surveillance to detect any spontaneous
heating and to observe any smoke. The hay must be kept in as
dry a state as possible because here, again, moisture serves to
accelerate bacterial action.

Bog and Swamp Areas and Compost Piles (Case E)

Bog and swampy areas, as well as compost piles, are all striking
examples of the susceptibility of decaying vegetable when covered and protected and possessing some moisture (not too much)
to accelerate the bacterial action.

I have had direct association with two particular examples,
which although different, were very similar in fire behavior.
The first of these examples concerned a wildlife reserve in
northern New Jersey which was administered by the US Department of Interior. This is a most luxuriant swampy area that
dates back to the last Glacial Period, and has plenty of standing water. The water table depends upon the frequency of rainfall. Rainfall in this area is usually not very different from
season to season. During the early and middle 1960s, however,
a drought occurred, causing the water table to drop, thereby
exposing (1) large areas of compacted decayed vegetable matter
resembling humus and (2) exposing the roots of trees that had
been covered with water but were now standing within an exposed "root ball." Fires began spontaneously within this root
ball, and tree after tree would then fall over. Ultimately the
large exposed areas of compacted vegetable matter would perhaps
become involved. A method of fire attack was performed on a
tree-by-tree basis using an aqueous solution of diammonium
phosphate, as previously described in Chapter 6. Penetration
of the root ball by this liquid stream was very effective, not
only for fire extinguishment but, of even greater importance,
for the fire-retardant coating that remained behind after the
water evaporated.

The second example concerns a very large-area swamp in
south Florida. The fire problem is caused by two factors: one
natural, and the other manmade. The ecosystem of the Florida
Everglades is such that the climate dictates wet and dry seasons that occur on a relatively regular basis. During the wet
season, which normally occurs in spring and fall, the area is
ordinarily covered with a shallow covering of water. The original system drained the overflow of Lake Okeechobee to the
south into the Gulf of Mexico. With heavy rains added to this,

the water table was usually higher than ground level. This resulted in vast deposits of rich "muck" south of the Lake Okeechobee area. Within the last half century, these areas became populated, and man has attempted to harness this system by installing a vast network of dikes and drainage canals. Hence, the deposit of additional layers on the muck base was stopped. Drainage and cultivation caused further lowering of the water table to a point that normal rainfall now does not always bring the water table up to provide moisture to the soil. In the dry season, which occurs winter and summer, the area becomes relatively dry although still moist enough to accelerate the bacterial activity into causing spontaneous combustion.

The fertile area surrounding Lake Okeechobee is cultivated extensively for the growth of sugar cane. After the harvest, the cane fields are usually burned down to clean out the trash and debris accumulated by the growth of the cane. The periodic cultivation keeps the soil subject to being disturbed, thereby preventing the slow accumulation of the developed heat resulting from oxidation. However, in the more remote areas that are not cultivated, fires have become serious problems particularly because of the inaccessibility of the area. Fires are usually fought by crews of the Florida State Forestry Service, the National Park Service of the US Department of Interior, and local fire departments. Fire crews are brought in by helicopter to fight the fire by hand or, where possible, to use bulldozers to plow fire lines and isolating burning areas.

Uniquely, before man's knowledge of fire prevention, nature would periodically burn the areas, thus cleaning debris on a regular basis. When fire control became a factor, it became apparent that extinguishment would only be temporary because of the inevitable resumption of spontaneous combustion in some other location as a result of the tremendous accumulation of decayed vegetation from centuries past.

Rotating Automobile Tires (Case F)

Case F, as illustrated in Figure 7.1, presents an entirely different aspect of the heat generation needed to incite spontaneous combustion. It is a matter of concern because of tire fires and the accidents possibly caused by them on the nation's highways. This situation has been highlighted by the large segment of interstate and interurban highway freight traffic carried on trucks moving at high, sustained speeds on the nation's high-

way systems. Many cargoes are of a flammable, toxic, or explosive nature, with close proximity to the public.

   To explain and define the problem, it is best to look at some
basic facts. Refer to Figure 7.3, that depicts the relationships
of stress versus strain for steel and vulcanized rubber. Metallic crystalline structures, such as aluminum, copper and, particularly, steel, exhibit a constant relation of stress to strain
within the limits of elastic proportionality. For instance, as a
spring becomes stretched, it develops a greater resistance
force and, conversely, when the spring is relaxed, the stretch-
ing forces diminishes proportionally. When, for a given-sized
test specimen, the strain is expressed as elongation and the
stress is expressed as a load, the area under the straight line
plot represents energy. As the spring is stretched, strain energy is imparted to the sample and, conversely, as the spring
is relaxed, the strain energy is returned from the sample as
illustrated on the left side of Figure 7.3. This phenomenon was
first formulated by the British mathematician and physicist
Robert Hooke (1635–1703), after whom the phenomenon was ap-
propriately called Hooke's law. The description is admittedly
idealized, and technically speaking, the returned energy is mi-
nutely less than the input energy, causing a barely detectable
temperature rise. It has been too small to even be considered.

   However, rubber compositions are amorphous and can be
made to vary over wide ranges of elasticity by chemical modification. Shown on the right side of Figure 7.3 is an approximate example of a vulcanized rubber. The exact shapes of the
loading and the relaxation curves are not of importance, at this
point, except to illustrate a fundamental behavior of vulcanized
rubber. At no portion of these curves is the stress proportional to the strain but, nevertheless, vulcanized rubber is
elastic in the sense that when the load is removed, it is re-
stored to its original length. On decreasing the load, the
stress versus strain curve is *not* retraced. The lack of coin-
cidence of the curves for increasing and decreasing stress is
known as "elastic hysteresis." Therefore, the area bounded by
the two curves is equal to the energy dissipated within the material, which is in the form of heat energy. The degree of energy transference can be controlled by chemical means employed
in the preparation of the rubber batch. This property of energy transference is of value in partially isolating the effects
of vibrating machinery from foundations. The heat energy developed manifests itself as a rise in temperature.

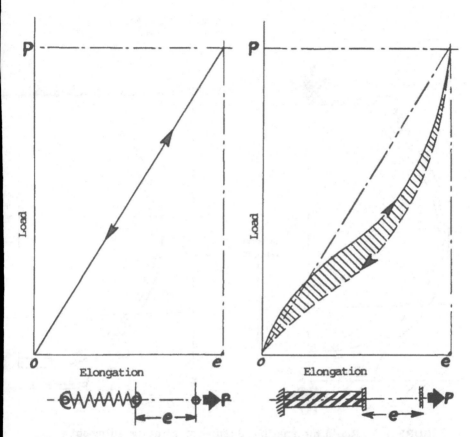

FIGURE 7.3 Typical elastic hystersis behavior of a vulcanized
rubber test sample compared with a steel spring of comparable
average elasticity. The cross-hatched area between the load
application and load relaxation curves represents the energy
converted into heat.

Figure 7.4 shows a particular embodiment of this application
wherein the automobile tire is inflated such that the relatively
thin walls of the tire carcass are prestressed to enable the tire
to support a load with the assistance of a cord or steel wire
reinforcement embedded therein. This load causes the tire to
compress without buckling as shown on the right side of Figure
7.4. Although all of the foregoing is obvious, it must be realized that the tire is cyclically loaded and unloaded for every

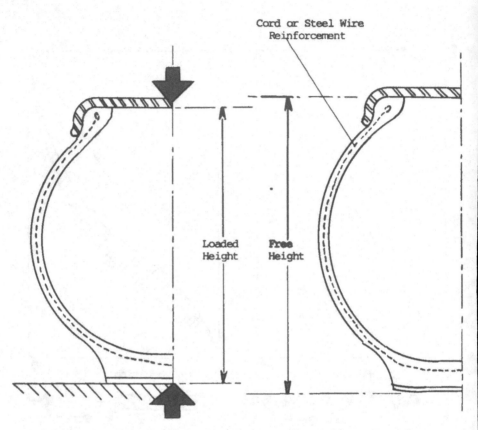

FIGURE 7.4 Rapid alternating flexure of rotating automobile
tire as a source of heat (see Fig. 7.3)

revolution at a frequency dependent upon the speed and the
nominal outside diameter of the tire. A truck tire having an
outside circumference of 10 ft will rotate 570 rpm (9.5 cps) at
a vehicle speed of 65 mph. A small unbalance can add significantly. The vibration of the tire is partly damped by the rubber tire. Although the flexure is very different from that presented in Figure 7.3, the same elastic hysteresis occurs, and
so evidently heat energy is being generated by the rotating
system. The tire temperature will mount until an equilibrium
condition is reached between the rate of heat energy developed
and the rate of heat energy dissipated. As the tire tempera-

ture increases, the pressure of the inflation air also increases. If the tire was "underinflated" to start with, the flexure and, hence, the rate of heat dissipation would have increased correspondingly. This would result in higher tire temperatures with a corresponding increase in inflation air pressure. The tire temperatures reached are subject to many variables, such as load, speed, tire diameter, inflation pressure, and weather conditions, such as temperatures ranging from desert heat to rain, ice, and snow. The problem is further exacerbated by the fact that rubber compounds are poor conductors of heat, which results in a further increase in temperature.

An underinflated tire becomes visibly obvious only when stationary. When in motion, however, as previously described, the inflation pressure is artificially raised, the increasing temperature and the bulging tire carcass is not apparent. Under proper load, pressure, and speed considerations, tires can even then become so heated that they are noticeably hot to touch with bare hands.

At about 420°F (215°C) the bonding of the rubber matrix to either the reinforcing cord or steel wire plys disintegrates and internal combustion starts under the urging of the inflation air. Externally, the tire is not yet visibly flaming, but there is a noticeable bluish smoke emanating from the tire. Even if this condition is noticed soon enough and the vehicle is brought to a full stop, the spontaneous combustion having started does not cease; the damage has been done and air under pressure is feeding the fire. To approach the smoldering tire is dangerous because such tires have been known to explode through the side walls. The only way to prevent this from happening is to regularly check the tires and note if a particular one (or more) is excessively hot when compared with the remaining tires.

An internal fire, such as described, cannot be successfully extinguished short of total immersion in water or high-expansion foam. These conclusions were reached by tests conducted by the US Army at its proving grounds in Yuma, Arizona. Loaded trucks with underinflated tires were deliberately run in the heat of the desert at various speeds to destruction, and conventional fire-fighting tactics were futile.

In the event of imminent danger and before the occurrence of smoke, the usual procedure is to proceed at low speed to a service area. The bleeding off of inflation air is a procedure that must be carried out with caution because of the danger of the fire-weakened tire bursting, scattering burning rubber tire shreds in all directions. There is a final note of interest, relative to tire fires, that is quite significant. It concerns the inflation of tires with pure nitrogen gas instead of air. The natural-heating process of the tire is not altered in any way. Because under this condition, no internal fire exists, the heating process will persist until ultimate collapse. This procedure has been effectively used on automobile race tracks and for the landing wheels of aircraft.

The obvious reason for emphasizing these hazards accompanying the driving of heavy, fast-moving highway trucks is that they are of concern in addition to all of the other hazards that are present. Another factor that makes this element of such importance is the close interface that exists with the traveling public using the same highways in open country as well as in congested areas, interchanges, bridges, vehicular tunnels, and the like.

ENVIRONMENTAL CONSIDERATIONS BECAUSE OF EXPOSURE TO RADIATION FROM OPEN FIRES

Another important environmental consideration is due to the transfer of heat to surrounding exposures, material as well as human, that result from the unimpeded radiation from fires. Upon turning back to the Frontispiece and to Figure 1.4, in particular, we recall that the text has frequently referred to the radiative feedback that provides the prime stimulus for the fire in conjunction with the other basic requirements, such as air and the nature of the fuel. The total amount of heat released, as shown and described in Figures 7.5, 7.6, and Table 7.3, is a composite of the convective heat flux, which is in the form of rapidly rising effluent gases and smoke, as well as the direct flame radiation itself, which is omnidirectional. Table 7.3 is a listing of 23 various common flammable liquid chemicals that are categorized into five basic classes, plus three extra miscellaneous ones, together with their respective heats of combustion and vaporization in BTU per pound. From this data, the values of the dimensionless parameter H_c/H_v were determined. From the method shown in Chapter 1, for radiative feedback, Figure 7.5 was obtained, wherein burn-off rates, measured as inches per minute and gallons per minute for each square foot of pan

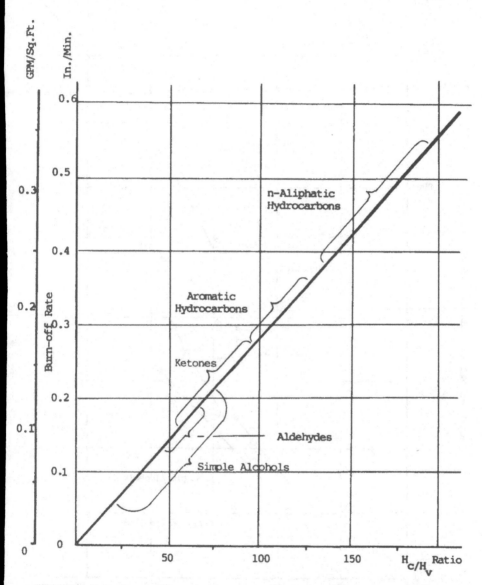

FIGURE 7.5 Burnoff rate of some liquid flammables versus the H_c/H_v ratio for circular fire pans in excess of 3-ft diameter. Fuel burndown rate (in./min) = (0.0028) H_c/H_v (see Fig. 1.4 for burning conditions).

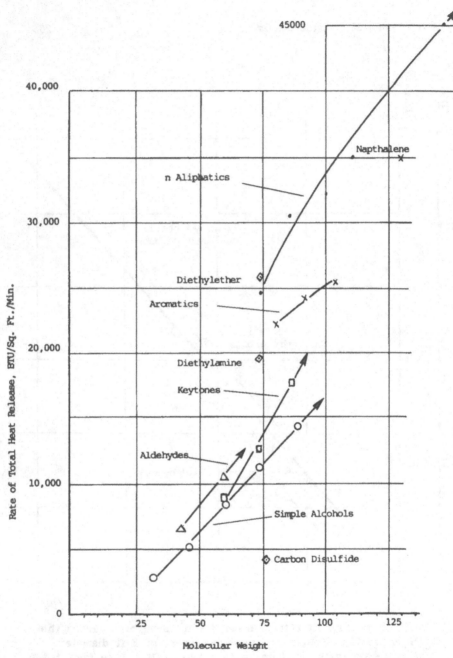

FIGURE 7.6 Rate of total heat release for various common liquid flammable chemicals versus molecular weight (square or round pans, 10 ft², or more).

TABLE 7.2 Generalized Summary of Types and Causes of Spontaneous Combustion

Cause	Oxidation					Elastic hysteresis
	A[a]	B[a]	C[a]	D[a]	E[a]	F[a]
Susceptibility because of unsaturated oils and/or fats plus a small amount of moisture	X		X			
Susceptibility because of organic and inorganic sulfides plus a small amount of moisture		X	?			
Susceptible to bacterial action plus a small amount of moisture	X		X	X	X	
Rapid alternating flexure						X

[a]Case number (see Fig. 7.1).

area are shown as functions of the ratio H_c/H_v. We can immediately detect that the five classes of flammable liquid chemicals become grouped into families.

Now, by reverting to Table 7.3 and continuing the relationship between molecular weight, specific gravity, and gravimetric burn-off rate and the heat of combustion for each of the chemicals, we can depict their characteristic rate of heat release properties measured as BTU per minute for each square foot as a function of molecular weight as well.

Very obviously, the n-aliphatic compounds are the greatest heat-releasing fuels. This idiosyncrasy is of significance and n-heptane, because of its general similarity to burning behavior of gasoline, was chosen by the Underwriters' Laboratories, as the standard test fuel. There is another reason for this choice: n-heptane is a pure liquid and not a mixture of normal and *iso*-aliphatic compounds, as illustrated in Figure 3.4. It has the

TABLE 7.3 Determination of Total Heat Release Rate of Various Common Liquid Flammable Chemicals

Class	Name	Formula	Mol wt	Heat of combustion (H_c)[a]	Heat of vaporization (H_v)[a]	H_c/H_v	Sp gr	gal/ (min·ft²)	lb/ (min·ft²)	Rate of total evolved heat [BTU/ (min·ft²)][b]
Simple alcohols	Methyl	CH_3OH	32	9.67	0.48	20	0.8	0.033	0.220	2.13
	Ethyl	C_2H_5OH	46	12.91	0.37	35	0.8	0.059	0.392	5.07
	Propyl	C_3H_7OH	60	14.60	0.30	49	0.8	0.084	0.559	8.12
	Butyl	C_4H_9OH	74	15.63	0.26	61	0.8	0.106	0.706	11.03
	Amyl	$C_5H_{11}OH$	88	16.33	0.22	75	0.8	0.132	0.879	14.36
Aldehydes	Acet-	CH_3CHO	44	11.56	0.24	47	0.8	0.081	0.540	6.24
	Propion-	CH_3CH_2CHO	58	13.40	0.20	67	0.8	0.116	0.773	10.36
Ketones	Acetone	CH_3COCH_3	58	13.23	0.24	55	0.8	0.094	0.626	8.28
	Methyl ethyl	$C_2H_5COCH_3$	72	14.54	0.19	76	0.8	0.132	0.879	12.78
	Diethyl	$C_2H_5COC_2H_5$	86	15.33	0.16	94	0.8	0.172	1.146	17.57

Aromatics	Benzene	C_6H_6	78	18.03	0.193	93	0.9	0.162	1.215	21.90
	Toluene	$C_6H_5CH_3$	92	18.26	0.177	103	0.9	0.178	1.330	24.28
	Xylene	$C_6H_4(CH_3)_2$	106	18.51	0.176	105	0.9	0.182	1.363	25.23
	Napthalene	$C_{10}H_8$	128	17.31	0.136	127	0.9	0.220	2.016	34.89
n-Aliphatics	Pentane	C_5H_{12}	72	20.94	0.154	136	0.6	0.237	1.185	24.82
	Hexane	C_6H_{14}	86	20.84	0.145	144	0.7	0.250	1.458	30.34
	Heptane	C_7H_{16}	100	20.82	0.138	151	0.7	0.263	1.534	31.93
	Octane	C_8H_{18}	114	20.69	0.132	157	0.7	0.275	1.667	34.50
	Nonane	C_9H_{20}	128				0.8			
	Decane	$C_{10}H_{22}$	142	20.52	0.109	188	0.8	0.331	2.204	45.27
Miscellaneous	Diethylamine	$(C_2H_5)_2NH$	73	17.87	0.164	109	0.7	0.188	1.096	19.60
	Diethyl ether	$(C_2H_5)_2O$	74	15.78	0.098	161	0.7	0.281	1.639	25.86
	Carbon di-sulfide	CS_2	76	5.85	0.151	39	1.3	0.066	0.715	4.18

aThousands of BTU/lb.
bThousands of BTU/(min·ft2).

main advantage that successive fires can be easily and quickly refueled, thereby not requiring the emptying of fire pans after every two test fires.

Note that in the listing shown in Table 7.3, n-decane has the highest value for the rate of total evolved heat per square foot. If this table had been carried on farther, even higher rates would be obtained with fuels such as heavy bunker-type fuel oils, asphalts, and napthenic constituents of coal tar.

An interesting confirmation was reported by the Sandia Laboratories in New Mexico from a series of tests designed to study the heat-resistant behavior of aircraft bombs that might become involved in an aircraft crash landing. When comparing the total heat release rates of n-heptane versus n-decane (the prime constituent of aircraft jet fuel) the latter showed a significantly higher rate of total heat release. When viewing Figures 7.5, 7.6, and Table 7.3, the reader must be reminded of the constraints of the test procedure that was established by the US Bureau of Mines and illustrated in Figure 1.4. To remind the reader, the basic conditions were

1. Use of circular (or square) fire pans exceeding 10 ft^2 in area
2. Filled brimming full of fuel
3. Outdoor environment (diffusion flames)
4. Dead calm
5. Standard atmospheric conditions

Hence, the only variable is the type of flammable liquid fuel being tested. The term "radiation" describes the undiminished transfer of energy through space unless its free passage is hindered by some material (solid, liquid, or gaseous) that either transmits, absorbs, or reflects this energy. The energy is electromagnetic and is transmitted in the form of waves, just as light is, traveling in straight lines at the velocity of light. The modern view of electromagnetic radiation is dualistic in that it has both the properties of waves as well as so-called corpuscle. The phenomenon of light propagation is best explained by the wave theory, whereas the interaction of waves with matter, in the process of absorption or emission is a corpuscular phenomenon, as for example, the heating effect of infrared rays, the initiation of chemical reactions, and photoelectric emission, all of which are explainable by the quantum theory developed by Max Planck. The heart of this theory is the direct relationship between wave frequency and energy. In the field of fire phenomena, our interest is concentrated on a small part of the entire electromagnetic spectrum, namely, the infrared, visible, and ultraviolet radiations that are intimately associated with temperature. Radiant heat flux is measured by means of a bolometer, an instrument based upon measurement of the electrical resistance of a blackened metallic conductor.

Because of the virtually infinite types of fires, it is necessary to confine this presentation to a simplified scenario involving a moderately sized outside fire with the fuel being a flammable liquid within a square fire pan with free access of air to all sides. In addition to this initial constraint, let us make the following assumptions to estimate the overall fire output:

1. Use n-heptane as the liquid fuel.
   From Table 7.3

   $$\frac{H_c}{H_v} = 151$$

   Fuel consumption = 1.53 lb (0.263 gal)/(min·ft^2)
   Total heat evolved = 31,900 BTU/(min·ft^2)
2. Because of incomplete combustion, we can probably assume about 13 lb of air will be needed to burn 1 lb of fuel.
3. There is ample data to substantiate that flame temperatures can be expected to range from 2500 to 2800°F (1371 to 1538°C).
4. We can assume, with assurance, that the specific heat content of the heated effluent gases and smoke is approximately equal to 0.32 BTU/(lb·°F) rise in temperature.

When all of the foregoing factors and assumptions are combined, the result will express the distribution of the total heat evolved and, as previously stated, will be (1) a composite of convection heat flux in the form of rapidly rising effluent gases and smoke and (2) the radiative heat flux emanating from the flames in an omnidirectional manner. This distribution, for diffusion flames from fire pans 10 ft^2 and larger, develops as two-thirds being shared in the form of convective heat flux and one-third being in the form of direct flame radiation. The aggregate total ra-

diative heat flux will be approximately 10,000 to 13,000 BTU/
(min·ft^2) of fire pan area. Radiance is measured in various
units, namely as

    BTU/(min·ft^2)
    W/cm^2
    cal/(sec·cm^2)

The equivalence of these units is as follows:

    1 W/cm^2 is equivalent to 52.5 BTU/(min·ft^2)
    1 W/cm^2 is equivalent to 0.24 cal/(sec·cm^2)
    hence, 1 cal/(sec·cm^2) is equivalent to 220 BTU/(min·ft^2)

In discussing radiance a point of reference is that solar radiation in temperate climes averages 0.10 W/cm^2 at noontime in
bright sunlight. It is a radiance level that human beings have
become comfortable with and is a good level to proceed to higher levels of radiance. A compendium of various and assorted
test data relating lateral ground level radiance as a function of
the fire pan size, and subject to all of the aforementioned constraints, is illustrated in Figures 7.7 and 7.8. Again, upon
referring to Figure 7.7, we can see that two types of exposure
hazards are tabulated. The first is the autoignition time interval, namely, from 1 to 3 min, in which six separate materials
will start to burn. Each material is accompanied by a nominal
radiance value expressed as BTU/(min·ft^2) and indicated on
the plot relative to the corresponding distance measured as a
multiple value of the fire "half-depth." The second type of exposure hazard is the response time for unprotected human beings. All of the foregoing data is predicated upon a radiance
value of 600 BTU/(min·ft^2) at the air-flame interface.

An interesting comparison can be made with two British papers that were delivered entitled "Radiation from Building Fires"
(1950) and "Heat Radiation from Fire and Building Separation"
(1963). It was mentioned therein that the radiant energy necessary to cause autoignition of unpainted, ovendried wood was
found to be about 0.3 cal/(cm^2·sec) for a period of 10 min.
From Figure 7.7, the same material was found to be about 175
BTU/(min·ft^2) [0.8 cal/(cm^2·sec)] for a period of from 1 to 3
min. The difference in the radiance values can be attributed
solely to the difference in time of exposure. Figure 7.8 presents the same data as Figure 7.7 but in a different manner,

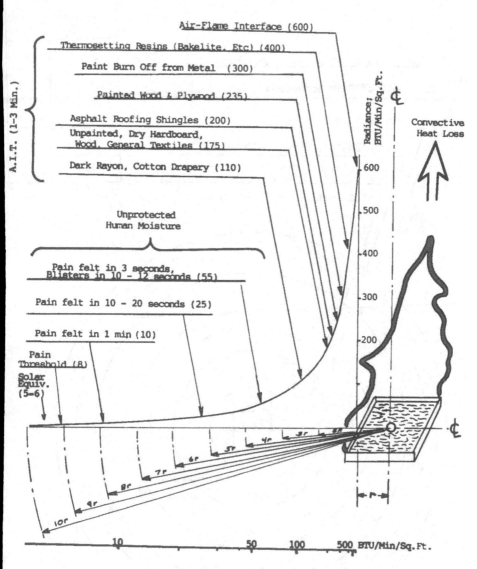

Lateral Radiance (Ground Level) As A
Function of Relative Distance From
Radiation Source

FIGURE 7.7 Ground-level exposure effects: circular or square fire pans or pools; hydrocarbon liquid fuel.

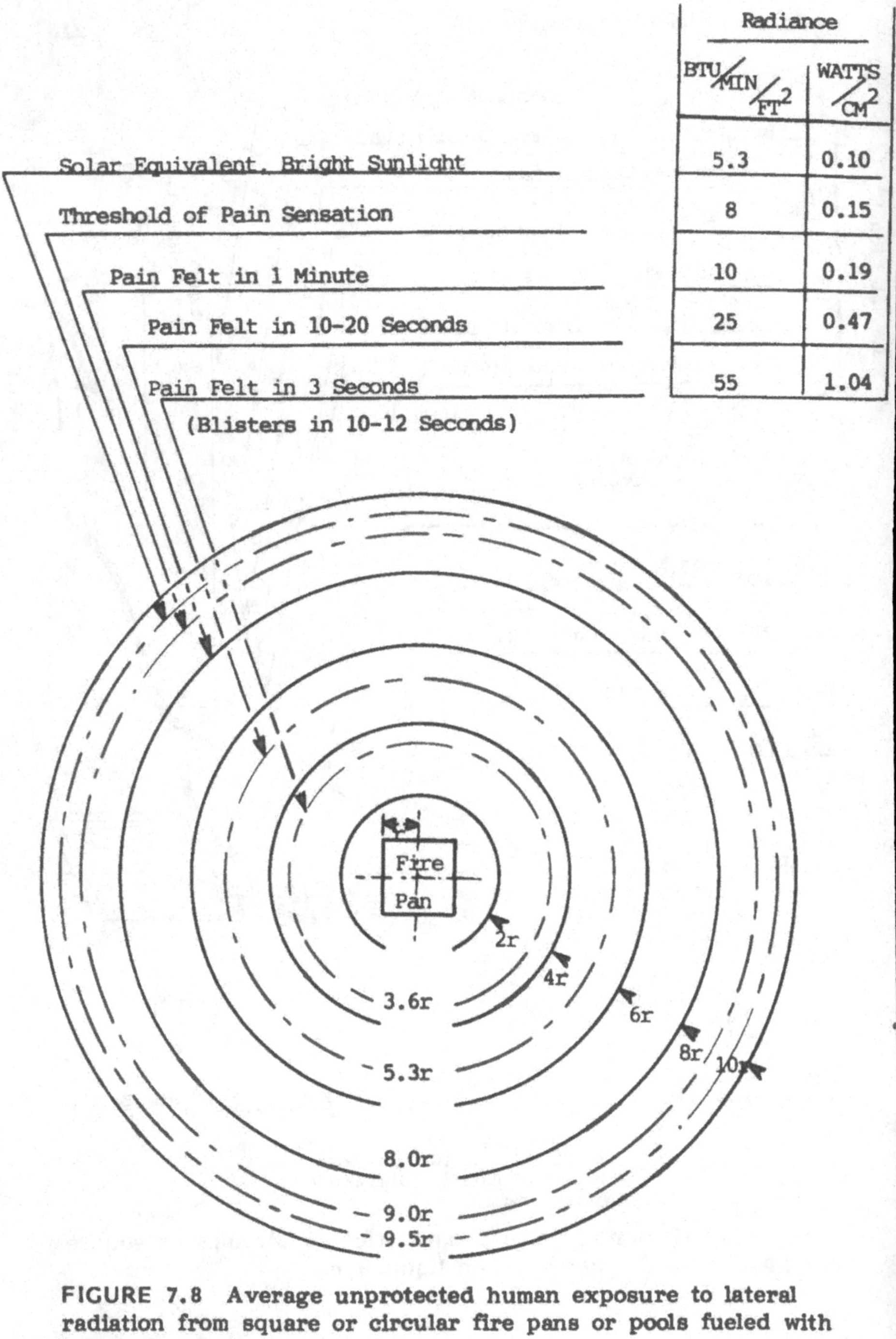

Radiance
BTU/MIN/FT²
WATTS/CM²
Solar Equivalent, Bright Sunlight
5.3
0.10
Threshold of Pain Sensation
8
0.15
Pain Felt in 1 Minute
10
0.19
Pain Felt in 10-20 Seconds
25
0.47
Pain Felt in 3 Seconds
(Blisters in 10-12 Seconds)
55
1.04
Fire
Pan
r
2r
4r
6r
8r
10r
3.6r
5.3r
8.0r
9.0r
9.5r

FIGURE 7.8 Average unprotected human exposure to lateral radiation from square or circular fire pans or pools fueled with *n*-heptane (ground-level).

concentrating upon the effects of radiance on human exposure.
To summarize, it can be said that the intensity of radiant energy at any point can be determined by

1. Its distance from the heat source expressed as a func-
    tion of the area or size of the heat source as seen from
    an exposed location. In Figure 7.7 this distance is ex-
    pressed at ground level. Table 7.4 is a tabulation of
    radiant exposure from a group of fires as measured at
    ground level.
2. The flame temperature. Thermal radiation intensity ema-
    nating from a radiating surface varies as the fourth pow-
    er of the absolute temperature; i.e., the radiant energy
    per unit area of a body at 1000°F (538°C) is about one-
    third that of a body at 1500°F (816°C) and only one-
    sixth that of a body at 2000°F (1093°C).
3. Maximum flame temperatures for the types of fires we
    are discussing can be expected to range from 2500°F to
    2800°F (1371°C to 1540°C). An optical pyrometer is the
    usual favorite instrument to use because of its conven-
    ience and because it does not require entry into the
    flame area.

ENVIRONMENTAL EFFECTS OF HEAT EMANATION, EFFLUENT GASES, AND SMOKE ON HUMANS

Human Adaptation to Heated Conditions

A considerable amount of research was conducted some years
ago, by the American Society of Heating and Ventilating Engineers, that showed the degree to which human adaptability could
cope with a heated environment. Because human adaptability is
a matter of great variability, it became necessary to narrow the
decision making to a healthy, young, normally clothed male.
The subject was tested at various temperatures between 60°F
(16°C) and 130°F (54°C), at an average 45% relative humidity
at each temperature level. During the test he exerted himself
at an average fixed rate at each temperature level of 1500 ft lb/
min (0.045 horsepower) for periods as shown at the top of Figure 7.9. Ample rest periods were allowed for recovery of normal pulse rate and body temperature, and for the relief of fatigue. The results of this extensive testing are summarized in

TABLE 7.4 Radiant Exposure As a Function of Fire Area, for a Group of Related Fires, Measured at Ground Level

Fire pan (ft)	10 × 10	20 × 20	40 × 40	80 × 80	160 × 160
Fire pan area (ft^2)	100	400	1,600	6,400	25,600
Fuel (lb/min)	153	612	2,448	9,972	39,168
Air (lb/min)	2000	8000	32,000	128,000	512,000
Air (ft^3/min)	28,000	112,000	448,000	1,792,000	7,168,000
Pan perimeter	40	80	160	320	640
Air [ft^3/(min·ft) perimeter]	700	1,400	2,800	5,600	11,200
Half-width of pan, r (ft)	5	10	20	40	80
AIT, 1.5 r 1–3 min. (ft)	7.5	15	30	60	120
AIT, 4 r 10 min (ft)	20	40	80	160	320
Radiance not above bright sunlight (ft)	50	100	200	400	800

} Unpainted wood

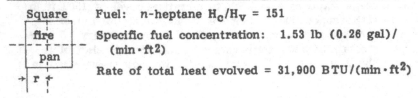

Square fire pan

r

Fuel: n-heptane $H_c/H_v = 151$

Specific fuel concentration: 1.53 lb (0.26 gal)/ (min·ft^2)

Rate of total heat evolved = 31,900 BTU/(min·ft^2)

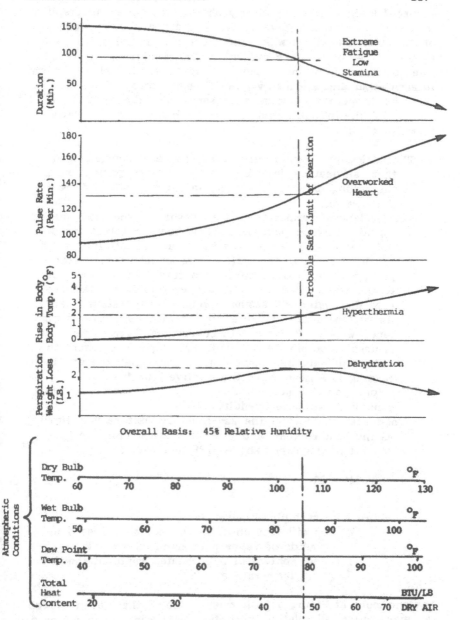

FIGURE 7.9   Human adaptation to heated conditions as described in text.

Figure 7.9, in which it is clearly shown that human adaptability
to heated conditions is determined by the prevailing temperature
of the enclosure within which the tests are carried out as well
as the degree of moisture present in the air during the tests.
The abscissae of the curve plots in Figure 7.9 are all equivalent
to each other and are involved with temperature. Some definitions for temperature become necessary. For the purpose of
Figure 7.9 the interrelationships are all based upon a relative
humidity of 45%.

   The "dry-bulb" temperature is a measure of sensible heat
       that can best be described as the sensory perception of
       the feeling of warmth. It is the reading obtained from a
       thermometer.
   The "wet-bulb" temperature is a measure of the total heat of
       the air—moisture mixture and, as such, involves the dry
       bulb temperature as well as the relative humidity, which
       can be defined as the ratio of the actual partial water va-
       por pressure to the saturation pressure at the dry bulb
       temperature. The wet bulb temperature is the tempera-
       ture that water will assume owing to evaporation in air if
       its surface area is large compared with its mass. A mois-
       tened wick or fabric sheath covering a thermometer bulb
       is used to obtain the wet-bulb reading. The extent to
       which the wet-bulb temperature is less than the dry-bulb
       reading is a measure of the relative humidity.
   The "dew point" temperature is a direct measure of moisture
       content or absolute humidity.
   "Total heat content" of the air—moisture mixture is defined
       as the heat content of dry air above 0°F plus the heat
       content of the vapor above 32°F and equals

       $$0.24t + Wh_S$$

       where t = dry-bulb temperature
             W = specific (or absolute) humidity expressed in
                 pounds of vapor per pound of dry air
           h_S = heat content of superheated vapor at the dry
                 bulb temperature.

As one looks at Figure 7.9, it must be noted that, even though
an average overall relative humidity of 45% was chosen for ease

of presentation, the dew point, as just defined, also increased
when the dry-bulb temperature was elevated

	Dry-bulb temp. (°F)	Wet-bulb temp. (°F)	Dew point temp. (°F)
Chilly	60	49	38
Temperate	70	57	48
Warm	80	65	57
Hot	90	73	66
Probable safe limit	100	81	75
Excessive	110	89	82

(All temperatures based upon 45% relative humidity.)

The metabolic complexes of physical and chemical processes within the human body that are involved in the maintenance of life
appear to impose certain restraints upon the performance of the
young subject, previously described, when working at a certain
rate of energy output.

From Figure 7.9 it would appear that under the assumed atmospheric conditions, the uninterrupted duration time should be
limited to those side effects, such as increased pulse rate, rise
in body temperature, and perspiration weight loss, all of which
occur under the indicated temperature conditions. Working beyond these constraints would lead to excessive fatigue, overworked heart action, hyperthermia, and dehydration. The extreme symptoms would be total exhaustion (excessive perspiration) leading toward excessive dehydration, which if not promptly attended can cause a heat stroke (syncope). Short-term exposures, at minimum levels of exertion can be tolerated at temperatures as high as 130°F (55°C) at relative humidities below
45%. Special ventilation provisions are needed for such locations as foundries, mines, and the like. The US Coast Guard
stipulates that the ventilation of manned machinery spaces in
merchant vessels shall be so designed as to limit the maximum
dry-bulb temperature, in all manned areas, to 120°F (50°C).

The metabolic output of heat energy of the body is shared
between (1) the respiratory process, after the assimilation of
oxygen and the consequent exhalation of heated air and carbon
dioxide; and (2) the perspiration process, followed by its evap-

oration to the outside air. The higher the dry-bulb temperature or the higher the humidity level, the more pronounced the difficulty of working in such an environment becomes.

Frequent and continuous exposure of workers to hot environments results in subtle physiological derangements affecting the white blood cells and other factors related to defense against infection. Another harmful effect is that the blood is diverted from the internal organs to the skin capillaries to assist in cooling. This diversion of blood carries with it the hazard of an anemic condition of the brain.

The physical fitness required of firemen would place them in approximately the same category as the hypothetical, average young healthy male described earlier. If any distinction is to be made, it is that the fire fighter has been trained in the required physical skills for effecting rescue and fire control. From a mental standpoint, he must also possess a basic knowledge of the many aspects of fire hazards to counter each situation with courage, tempered with wisdom. However, the fire fighter to perform his duties of rescue and combating the fire must, by necessity, be suitably clothed, including helmet, gloves, and boots, in addition to being equipped with self-contained breathing apparatus. All of these items are covered in detail in NFPA Standards 1971, 1972, 1973, and 1981.

It will be remembered that in Figure 7.9 we were referring to normally clothed, young, healthy males, whose metabolic functions were in good working order, while performing work. The only limitation was the uninterrupted duration of his efforts. However, with the suitably attired and equipped fire fighter described in the previous paragraph, his metabolic functions are working under both psychological and physiological handicaps. The only way that he can be expected to perform is to reduce the time factor. This is a purely judgmental matter. Figure 7.9 highlights the general conclusions about probable safe limits of exertion based upon the aforementioned tests that were conducted under the auspices of the American Society of Heating and Ventilating Engineers. If we can assume that safe limits of exertion still hold, then the fire fighter will have to drastically reduce his time of exposure. When the duration of the working man's exertion was limited to, say 100 min under the specified atmospheric conditions, the fire fighters time of exposure could very likely vary from 5 to 20 min. The hazards that come from overexposure could include low stamina (extreme fatigue); overworked heart (in excess of 130 beats/min); a rise in body temperature (in excess of 2°F), known as hyperthermia, and excessive perspiration, resulting in dehydration.

Another factor of importance is that under duress and exertion the breathing rate is increased considerably. Whereas a so-called "30-min," conventional, demand-type, self-contained breathing apparatus has sufficient air capacity to supply air under quiescent conditions for an average of 30 min, under heavy breathing and strong exertion, the period can be reduced by degrees ranging from 50 to 75%. Higher-capacity, self-contained breathing apparatus, as well as newer types of rebreathing apparatus, are available with the purpose of extending this time with as little weight penalty as possible. The extra time must be tempered with the full realization of the physiological limitations illustrated in Figure 7.9.

As basic and fundamental as radiance levels are, they have meaning only when coupled with a time factor. The heat-protective attire required by fire fighters has been the subject of intensive study to establish criteria for its combined radiation resistance, water resistance, and abrasive wearing. Much research in this area has been carried out by the government, by trade associations, and by private companies, and there has been much controversy. Collective agreement has not generally been reached. To attempt to reach a consensus, the NFPA Standard 1971 employs three methods to establish safety criteria.

1. An ignition test utilizing an adjusted 1550°F (843°C) flame tip impinging for a period of 12-sec exposure at the bottom of a vertically positioned sample. Maximum char shall not exceed 4 in. in length. Flaming is allowed only during the 12-sec period and must cease when flame is removed.
2. Heat stability is determined by a 5-min exposure to the outlet from a special forced-air oven, having a 50:50 ratio of convective and conducted heat. The sample is exposed to 482°F (250°C) and must not char, separate, or melt.
3. A third test is conducted to determine actual heat transfer and relate it to the severity of burn injury. The test procedure is complex.

Each of these methods has a particular objective, and what remains to be achieved is to encompass them all into a practical combination. By the development of highly fluorinated nylon

textiles, remarkable flame resistance has been developed that
would permit very limited protection from AIT levels that are
capable of "torching" the various materials listed in the upper
portion of Figure 7.7 in a time frame of from 1 to 3 min. The
duration of this limited protection is very problematic, but it
surely would be within the shortened time duration of self-contained breathing apparatus when the fireman is exerting himself
and breathing heavily. The upper limit to which this line of
reasoning can extend is also problematic, but a radiance level
of about 300 BTU/(min·ft^2) or approximately 6 W/cm^2 can probably be sustained for a very short period (< 3 min). Closer
approaches to a fire can then be accomplished only by use of
aluminized apparel that, to be fully effective, must be kept in
an absolutely clean, soot-free condition to maximize the reflec-
tive properties of the aluminized gear. It is of great importance to eliminate the possibility of water intrusion in the event
that a hose stream impinges upon the clothing. The clothing
must be of a nonwater-absorbing material and one that maxi-
mizes resistance to tearing and abrasion.

The field of human protection from fire radiance is a highly
technical matter that will always be in a state of advancing improvement.

Smoke and Effluent Gases and Their
Effect on Unprotected Humans

When suddenly exposed to a fire, involuntary reactions induce
the danger of fear and its consequent mental stress because of
man's primeval "fight-or-flight" response to a threatening situation. Harmful physiological and psychological effects can become a problem. The intake of toxins is accelerated. It has
been shown that amphetamines, for example, are 10 times more
toxic under stress. Under stress the heart rate and blood
pressure both rise. Blood is diverted from the digestive system to muscles, the brain, and other organs. The production
of epinephrine and other hormones is increased, together with
the flow of fatty acids and glucose into the blood. The body's
reaction to infection is changed, and a lower toxic tolerance
threshold develops.

As will be shown, there are, at the very least, 13 hazardous
classifications of toxic substances involving fire. It is impossible to address these toxic ingredients individually because in
fire situations, in the aggregate, they can form fierce mixtures.

The individual toxic effects are not merely additive but can combine to form exceptionally harmful combinations. The individual threshold limit values (TLVs) for chemical substances and physical agents in the workroom environment have been tabulated by the American Conference of Governmental Industrial Hygenists, P. O. Box 1937, Cincinnati, Ohio 45201. Threshold limit values refer to time-weighted concentrations for a 7- or 8-hr workday and a 40-hr workweek. They are used as guides in the control of health hazards and not for fire lines to be drawn between "safe" and "dangerous" concentrations. The limits are based upon the best available information from industrial experience, experimental human and animal studies and, when possible, from a combination of all three. All of the foregoing is relevant to normally attired working people, performing their duties under normal atmospheric conditions. The presence of two or more of these noxious gases can be assessed by utilizing a method similar to that of the Le Chatelier procedure outlined in Chapter 3. It must be apparent that these types of data are of absolutely no use when applied to fire situations, except perhaps to qualitatively rate them in a loose manner according to their relatively toxic potential. Of the 13 generalized hazardous and toxic materials previously mentioned, we will list and describe only those of major importance.

Smoke

Incomplete combustion results in the formation of complex flammable tary substances that, upon thermal decomposition, develop into sooty molecular clusters that consist mostly of carbon. Besides serving to obscure vision and cause ocular irritation, both of which serve to cause disorientation, smoke can cause permanent damage to the lungs. Unless rescue is prompt, the victim is in jeopardy of being exposed for an excessive time to all of the following harmful contents of the fire effluent whose deleterious effects are stimulated by time and temperature.

A convenient measurement of smoke density is to observe the obscurant effect of smoke as shown in Figure 7.10. The optical system is adjusted for a parallel light beam to traverse a distance (L) within a cylindrical blackened (nonlight-reflective) tube through which air samples are continually introduced. The ratio of the light energy received by the voltaic photocell to the light energy emitted through the lens (I_2/I_1) is expressed as an exponential function in which the exponent contains the var-

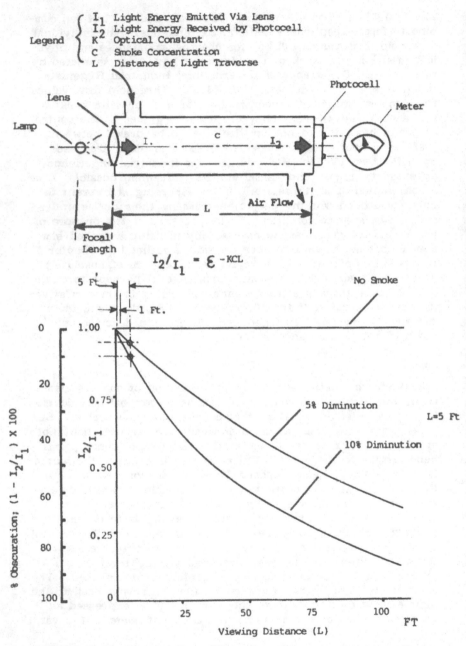

Legend
- I_1 Light Energy Emitted Via Lens
- I_2 Light Energy Received by Photocell
- K Optical Constant
- C Smoke Concentration
- L Distance of Light Traverse

$$I_2/I_1 = \varepsilon^{-KCL}$$

FIGURE 7.10 Obscurative effect of smoke upon the transmission of light as defined by the Beer-Lambert law.

iables of smoke concentration (c) and distance (L) and an optical constant (K). The exponent applies to the base of natural
logarithms (ε), which occurs frequently in problems concerned
with natural growth or decay, e.g., speed of chemical reactions,
decrease in atmospheric pressure with increasing altitude, radioactive decay, rate of cooling of a warm object. The smoke
is produced within an external smoke chamber by smoldering
pieces of "punk," prepared from vegetable fungi, that will smol-
der continually and not break out into open flame. The emitted
smoke is bluish grey. The air control to the smoke chamber
permits a constant smoke concentration to be produced and
maintained into an induced air stream. The flow is directed
through the measuring device and obscuration is measured from
the baseline of "no smoke."

A 5% obscurant effect within a 5-ft long tube is frequently
used as a measure and as a point of reference for either overheated smoking, flammable materials, or an actual fire. In Figure 7.10, it is interesting to note the reduction of visibility
when a light source is viewed from various distances in a closed
space when c is constant.

Distance (%)	Obscuration (%)
5	5
50	39
100	63
150	78 extrapolated
200	86 extrapolated

The preceding method of producing and measuring smoke levels
is a standard, not only for direct photoelectric smoke detectors
but also for photoelectric smoke detectors working on the principle of forward light scattering with a dark background. The
Fire Protection Handbook provides a source for detailed information on all of the various types of smoke detectors.

Carbon Dioxide

During a fire, carbon dioxide is evolved in considerable quantity which, with interior fires, starts to accumulate and affect
the respiratory process. The metabolic system of inhaling air
and exhaling a mixture of air, carbon dioxide, and free mois-

ture is controlled automatically in response to the needs of the body. The breathing rate and tidal airflow are increased or decreased accordingly. However, the source of the involuntary motor system that controls the breathing rate and tidal flow is also affected by the carbon dioxide concentration. Normally, the carbon dioxide concentration in the atmosphere averages 0.033% over the entire world, but if the carbon dioxide concentration is raised to 2.0%, the rate and depth of breathing is increased 50% over normal conditions, and at 3.0% the increase amounts to 100%. This superventilation results in a correspondingly greater intake of toxic gases, which normally would pose enough of a threat. However, with a 100% increase in ventilation occasioned by the presence of 3.0% carbon dioxide, the story is entirely different.

Tests with humans, in particular with submarine crews, have shown that when the percentage of carbon dioxide exceeds 3.0% a reversal occurs and that, at 5.0%, physiological changes occur that significantly reduce the metabolic rate and efficiency of performance. The US Navy conducted dock-side tests in a submerged submarine in which the crew members were subjected to 5% CO_2 for 2 weeks. There were no harmful aftereffects. The threshold limit values (TLV) for carbon dioxide in a working environment has been set at 2.0% (5000 ppm). It has been claimed that the gas is nontoxic, but these statements must be tempered with the realization that as the carbon dioxide concentration is increased, the oxygen concentration of the atmospheric mixture is reduced. A mixture of 38% carbon dioxide and 62% air (13% oxygen) has been known to be lethal, producing a cyanotic condition that manifests itself as a disordered vascular circulation characterized by a livid bluish skin color. The low level of oxygen probably contributes as much of a lethal hazard as does the carbon dioxide.

Carbon Monoxide

Although not the most toxic of the various fire gases, carbon monoxide is in great abundance within burning buildings. Tests conducted by the Factory Mutual Laboratories showed that the atmosphere within burning buildings contained about 2 parts carbon monoxide for every 3 parts carbon dioxide. The reader is referred back to Figure 5.3 that depicts equilibrium mixtures of the two gases in the presence of incandescent carbon, which is characteristic of the yellow orange flames emanating from class A and class B combustibles (see Fig. 6.8). Figure 7.11 illustrates the relationship of potentially lethal carbon monoxide concentration to time of exposure. Both curves shown represent potentially lethal concentrations, one for quiescent, nonexertion conditions and the other for vigorous action involving heavy panting and exertion coupled with the superventilating action of accompanying carbon dioxide.

Carbon monoxide is generally formed by the combustion of complex plastic substances. Fire statistics have shown over and over that loss of life in building fires are caused by noxious mixtures of fire gases, of which carbon monoxide is the chief culprit. To complete this fateful situation is the presence of smoke. The effect is to first produce nausea followed by dizziness, disorientation, and muscular collapse, terminating in death, unless rescue is prompt. The peril that is peculiar to the human respiratory system is that the red hemoglobin content of the blood has approximately 200 times the affinity for carbon monoxide as it has for the oxygen in the air. The important role of metabolism is upset by robbing the blood of the oxygen (asphyxiation) in the air and forming carboxyhemoglobin. The complexion of the affected individual becomes very pink and flushed. If rescue is prompt, recovery can be accomplished with the help of oxygen inhalation and long rest. The maximum TLV in the workplace has been established as 50 ppm (0.005%), particularly in such locations as vehicular tunnels, bus stations, bus and auto maintenance shops, and the like.

Anoxia

Anoxia is a term that means simply a lack of oxygen. When oxygen drops from its usual sealevel concentration of 20.9% in air (3.07 psi partial pressure) to 15% in air (2.21 psi partial pressure) a person's muscular skill is diminished. As the oxygen percentage drops to 10% (1.47 psi partial pressure) a person is still conscious but has faulty judgment, which is not self-evident, and also becomes quickly fatigued. Below 10%, collapse becomes imminent and revival requires fresh air, or oxygen, or both. During periods of exertion, increased oxygen demand may result in a much higher percentage of oxygen deficiency symptoms.

In a fire situation, a deficiency of oxygen is accompanied by increases of other gases emanating from the fire, all of which are noxious. To discuss anoxia, by itself, may seem academic,

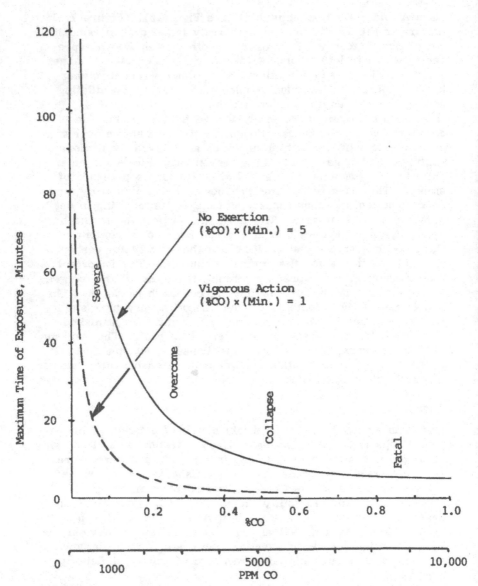

FIGURE 7.11 Physiological effects of time exposure to potential-ly lethal concentrations of carbon monoxide.

but it is of relevance when considering the human respiratory system. A deficiency of oxygen can be created in a number of ways. One is to be at an altitude where the partial pressures of atmospheric oxygen are lowered to the limits mentioned before.

2.21 psi partial pressure of O_2 (about 12,000 ft)
1.47 psi partial pressure of O_2 (about 24,000 ft)

These values are subject to variations and are given only as approximations. The need for auxiliary-breathing apparatus is most apparent.

Another way anoxia can be created is by the presence of some simple asphyxiants that are inert in physiological behavior and that are not detectable by odor. Their effect is not self-evident. Some are flammable and others are not. The following are some examples:

Flammable	Nonflammable
Hydrogen	Helium
Methane	Neon
Ethane	Argon
Propane	Krypton
Butane	Nitrous oxide
Ethylene	

The flammable gases, such as methane, ethane, propane, butane, ethylene, are the constituents of natural gas which, if not artificially odorized, will not be self-evident. If sufficiently strong concentrations exist to be within the limits of flammability, an explosion will occur, even though an asphyxiation condition does not exist.

Moisture

The burning of fuels of any type that contain hydrogen results in the formation of hydrogen oxide or free moisture. Whether or not moisture, by itself, is toxic is a moot point, but it does pose a jeopardy to human existence under the conditions of elevated temperatures. There are three aspects to be considered.

1. The smoke and gaseous effluent from hydrocarbon-bear-
    ing fuels contains a considerable amount of free moisture
    that imparts a distinct humidity to these gases. The
    amount has been variously quantified as a dew point of
    approximately 140°F (60°C). Considering that human me-
    tabolism is thermostatically controlled to temperature lev-
    els of 40°F lower (22°C lower) than this, the body, both
    internally and externally, becomes a condenser upon
    which this fire-produced moisture can condense, thereby
    surrendering its latent heat of vaporization and causing,
    by itself, an oppressive atmosphere that cannot be en-
    dured. The victim literally can drown under these cir-
    cumstances.
2. Certain gaseous constituents, such as ammonia, hydrogen
    chloride, and sulfur dioxide, are listed, respectively, as
    the most water-soluble gases. In this dissolved state,
    their toxic effects are enormously enhanced. Even at
    normal temperatures, all of these gases, which are solu-
    bilized in the moisture associated with eyes, mucous mem-
    branes, and perspiration, can cause acute irritation and
    harm.
3. The presence of moisture greatly accelerates many chem-
    ical reactions involving the gaseous constituents that
    evolve from fires, such as hydrogen sulfide, hydrogen
    cyanide, nitrogen dioxide, acrolein, phosgene, and oth-
    ers. The cross-mixtures that are possible complicate the
    problem even more. Quoting the TLV, as previously
    mentioned, becomes utterly meaningless.

The NFPA's *Fire Protection Handbook* is an excellent source for
more particular information.

Miscellaneous Other Toxic Gases

The origin of the various gaseous constituents, other than carbon dioxide, carbon monoxide, and moisture, lies within the nature of the fuel.

1. Fuels that contain chlorine, such as polyvinyl chloride,
    do burn with difficulty at low rates of heat emission but,
    as the rate increases, they burn with emission of hydro-
    gen chlorine and phosgene (a compound of carbon mon-
    oxide and chlorine, carbonyl chloride). Concentrations

of 0.15% can cause death in a few minutes. The gases are extremely pungent and irritating. Chemicals that are not flammable but contain chlorine in their structure, such as pentachlorophenol (PCP) and other chlorinated hydrocarbons, will decompose to release decomposition products similar to those of polyvinyl chloride, and they are extremely toxic.

2. The incomplete combustion of fuels that contain nitrogen, such as wool, silk, urethan acrylics, and polyamides, can emit substantial quantities of hydrogen cyanide. Exposure to 0.3% can be instantly fatal. This compound is a vermin fumigant and poses a serious problem in buildings that have been either fumigated or are used for storage. It can be absorbed through the skin.

3. The incomplete combustion of fuels containing nitrogen, such as wool, silk, acrylic plastics, and phenolic resins, can release ammonia. Its use in refrigeration and chemical-processing poses the problem of accidental release and its extreme solubility in moist mucous membranes, the moisture associated with eyes (causing blindness), and perspiration present terrible problems. Exposure to concentrations of 0.2 to 0.5% can cause death within 30 min.

4. Another emission from fuels that contain nitrogen, particularly cellulose nitrate, and from fires that involve ammonium nitrate, is the gas nitrogen dioxide (NO_2) or nitrogen tetroxide (N_2O_4). Either of these gases will dissolve in moisture, mucous membranes, perspiration, to form nitric acid. It is extremely toxic.

5. Hydrogen sulfide (H_2S) and sulfur dioxide (SO_2) are emitted from fuels containing sulfur, by contact of sulfides with sulfuric acid, and from certain metallurgical processes. Solubility in moisture is a problem, particularly so for sulfur dioxide which forms dilute sulfurous acid. Both gases are extremely poisonous.

Although an almost endless list of toxic chemicals can be listed, they are usually in small amounts and also are usually detected by eye irritation, pungency, or choking effects. The usual general main hazardous gases are carbon dioxide, carbon monoxide, and moisture that, together with smoke, have been statistically shown to be the major cause of fatalities in fire situations.

Epilogue

Knowledge is proud that
  he has learned so much;
Wisdom is humble that
  he knows no more.

        William Cowper

Education is the instruction of the intellect in the laws of
Nature under which name I include, not merely things and
their forces, but men and their ways, and the fashioning
of the affections and the will into an earnest and loving
desire to move in the harmony with those laws.

        Thomas Huxley

Whereas this text was devoted to the complex nature of fire,
the intricasies of the combustion process, the influence of chemical and physical properties of fuels, and the proper means of
fire extinguishment, it would be remiss not to place all of this
in proper perspective. The accompanying table (Table 1) presents the five phases of a fire hazard with the numerous items,

TABLE 1   The Five Phases of a Fire Hazard

1. *Prevention*

    observance of codes and standards

    good housekeeping

    periodic maintenance of all facilities

    periodic maintenance of fire protection apparatus

    periodic inspection by fire inspectors

    fire preplanning by fire officials

    evacuation drills—water supply

2. *Retardation*

    monitoring of unanticipated hazardous conditions

    alterations resulting from change of occupancy

    maintenance of exposure clearances

    flammable gas concentration detection

3. *Detection/alarm*

    manual (indeterminate time)

    automatic (0–2 min)

       fixed-temperature detector

       rate-of-temperature rise detector

       photoelectric optical smoke detector

       ionization products of combustion detector

       infrared radiation detector

       ultraviolet radiation detector

    local or remote alarm

4. *Control/extinguishment*

    response time by fire department[a]

    size-up and establishment of command post

    entry and rescue—external or internal attack

    selection of extinguishing agent(s) and tactics

    protection of exposures

TABLE 1 (Continued)

---

5. *Salvage/investigation*

   maintain fire watch

   search

   cleanup

   investigation

---

aGreatly affected by onsite automatic fire extinguisher system.

the attention to which can mean yes or no to the outbreak of a fire. Perhaps, nowhere can this matter be more forcibly exemplified than in the example of a ship at sea. It is a community unto itself, self-contained and dependent for its safety on a small group of officers and crew to run a "good tight ship" under all conditions.

The *first* item of the tabulation is prevention. This is a never-ending task that requires the services of knowledgeable fire inspectors who have a strong background in codes and standards coupled with native good judgment (commonly called common sense) practiced in an uncommonly thorough and observant manner. *Prescience* becomes a valuable asset. If the various items listed under the heading of prevention are in order, we can say that the operation or situation is safe ("green light" or go signal).

The *second* item of the tabulation is termed "retardation" and is, in effect an indication that something is amiss that could jeopardize safe operation. Many examples occur, such as overheated flues, overloaded electric circuits, unusual vibratory conditions, poor pressure regulation, placing combustibles too close to furnaces and heaters, excessive flammable gas concentrations, and many others. The usual fire inspector is not generally present to observe this, but properly trained personnel under the guidance of a safety-minded management should be alert and follow proper procedures. Too often deficiencies are blamed upon either too few or poorly trained personnel but management must assume all responsibilities. The imminence of the outbreak of fire is difficult to foretell with any degree of exactitude; therefore, we must take corrective measures ("yellow light" or caution signal).

The *third* item of the tabulation is of extreme importance in that it is involved with the time interval between the first recognition of fire (either human or automatic) and the time of response. Both of these time factors are subject to considerable variation.

1. The manual alarm will be recognized as the usual wall-mounted alarm box that is manually actuated by someone who has discovered open flames and smoke. The unknown factor is the length of time since a fire erupted.

2. The automatic fixed-temperature detector is available in two versions: a nonelectric and an electric assembly that operate by either the fusing of a specific low-melting temperature-fusible alloy or by a chemical compound to actuate a sprinkler head, mechanical or electric alarm, a spring-tensioned lever that activates a valve, releases a fire door, or some other such system. This type of detector is the slowest to respond to heated conditions. For ceiling-mounted automatic sprinkler heads, actuation time can vary from 0 to 2 min, when they perform according to Underwriters' Laboratories or Factory Mutual Laboratory Standards.

3. The automatic rate-of-temperature rise detector is also available in a nonelectric, as well as an electric, version. Instead of responding at a fixed-temperature level, responds to 15°F (8°C)/min rise in temperature within a closed space. Such a detector would detect an unnatural rise in temperature except in the vicinity of oven, furnaces, metallurgical operations, and such.

4. The photoelectric smoke detector, as previously mentioned in Chapter 7, is tested and standardized on the basis of a bluish grey smoke. Heavy black smoke would render the detector more sensitive. Whereas the detector illustrated in Figure 7.10 functions on the basis of light obscuration of the path of a light beam, another detector that is widely used detects refracted light against a dark background, that was derived from light scatter either to the side or forward from the main light beam. Smoke particle sizes, to affect obscuration, must be approximately 1 to 2 μm.

5. The ionization product combustion detector, which is a modification of the well-known Wheatstone bridge principle, compares the electrical conductivity of two adjacent chambers, both of which are subjected to alpha-rays of a very weak source of radioactivity. One of these chambers is not subject to the air sample being measured, whereas the other chamber is the one

through which the air sample is being passed. The electrical values are extremely minute and utilize a cold cathode tube for detection. It is reported that the invisible particles of smoke (if that is what we can call them) range in size from 0.01 to 1.0 μm, which would be far larger than what we conceive to be individual molecules. Apparently what occurs is that the freshly formed heated gases from fires are not formed individually, but as clusters, and in this macromolecular condition can interfere with the action of the alpha-rays being emitted in the detector and, thus, produce an effect. It is also noteworthy that, as the smoke "ages" or as gases slowly arise from smoldering fires, this clustering appears to be reduced. The clustering can be attributed only to electrical charges upon the molecules which occur during or just following the combustion process.

Note: It must be stated that all of the foregoing types of fire detectors are subject to air movement throughout the enclosure. Usually, it is considered that the time interval following ignition should not exceed 2 min. This is based upon many observations of the growth of fire.

6. The infrared and ultraviolet fire detectors are highly sophisticated devices that have distinctive features for separating flame radiations from accompanying similar (but different) characteristics. For instance, for the infrared flame detector, the reader is referred to Figure 4.7; the flame flicker radiation shown can be selectively detected by electronic means. The detector will not respond to sunlight, incandescent light bulbs (60 cps), or heated equipment, such as the exterior of a furnace or oven. It will respond only to infrared radiations emanating from open flames. The detector does not have to be aimed at any preselected area because the infrared radiations are reflexive, just as light rays are. Furthermore, the detector has no thermal lag and is not subject to stray air currents. The ultraviolet flame detector also is equipped with the distinguishing ability to separate the ultraviolet radiations of flames from the ultraviolet portion of sunlight. Both the infrared and ultraviolet fire detectors are capable of very rapid detection.

Note: An excellent and detailed description of all the various types and variations are contained within References 1 and 5 of the Suggested Reading list.

The *fourth* item in the tabulation begins with the vital matter of response. The fire-controlled exothermic nature of a fire causes a rapid evolution action that demands prompt detection and rapid response. From that point on fire fighting be-

comes a very hard practical matter, requiring the ability to adjust to changing conditions.
Note: An excellent text on basic fire fighting techniques is contained in Reference 2 of the Suggested Reading list.

In 1973, the National Commission on Fire Prevention and Control submitted its final report entitled "America Burning," in which the fire problem in the United States was defined and put into proper perspective. A most valuable excerpt from this report concludes as follows:

> Indifferent to fire as a national problem, Americans are similarly careless about fire as a personal threat. It takes the careless or unwise action of a human being, in most cases, to begin a destructive fire. In their home environments, Americans live their daily lives amid flammable materials, close to potential sources of ignition. Though Americans are aroused to issues of safety in consumer products, fire safety is not one of their prime concerns. And often when fire strikes, ignorance of what to do leads to panic behavior and aggravation of the hazards rather than to a successful escape. The fire problem must be addressed as a personal threat: personal behavior before and during fire emergencies must be changed so that individuals will know how to prevent fires, protect themselves in the event a fire does occur and to persuade others to prevent and protect themselves from fire.

The NFPA Standard 901 entitled "Uniform Coding for Fire Protection" is the source for the following definitions of fire casualties:

1. A fire casualty is a person receiving an injury or death resulting from a fire. The cause may be either *direct*, a result of the fire, or *indirect* when it is only partly a result of fire, but some other cause is considered as being primarily responsible.
2. Fire injury is one that requires medical treatment within 1 year after the fire.
3. Fire death is one that is immediate or one that occurs within 1 year after the fire.
4. When a death may be due to more than one cause, one of which is fire, the classification made by appropriate authorities shall govern provided this classification is in accord with the *Manual of the International Statistical*

Classification of Diseases, Injuries and Causes of Death (PHS Publication 1693, Superintendent of Documents, Government Printing Office, Washington, DC 20402).

It is not difficult to realize that statistics obtained from the foregoing classifications will show discrepancies that are usually the result of lack of agreement about what constitutes a particular fire casualty. It is a sad commentary that a high annual fire death and injury rate appears to be peculiarly an American problem. No other industrialized, free nation even comes close as shown in Table 2. As ragged as this tabulation is, it does unequivocally show that the United States leads the list by a

TABLE 2 Fire-Related Deaths and Injuries by Country Between 1970 and 1980

Country	Deaths/million	Injuries/million
United States	52	620
Canada	35	N/A
Australia	31	214
Belgium	28	105
United Kingdom	28	N/A
New Zealand	18	120
Japan	17	83
Finland	16	N/A
Norway	15	N/A
Sweden	12	N/A
Denmark	12	N/A
France	8	42
Italy	8	20
Netherlands	6	42
Switzerland	3	42

large margin. The reasons that this should be so would, no doubt, be a matter of philosophical speculation. The following list of possible reasons are presented, but they are not necessarily in any order of importance.

1. Highest standard of living
2. Life-style
3. Public apathy
4. Demographical
5. Sociological
6. Political
7. Conflicting rules and regulations

On the other hand, it is ironic to note that most of the research, design, and development in the fields of fire combat and fire detection have been contributed by the United States. Our task is clear. There is no substitute for fire prevention.

Suggested Reading

Bryan, J. L. (1974). *Fire Suppression and Detection Systems*. Glencoe Press.

Clark, W. E. (1974). *Fire Fighting Principles and Practices*. Dun-Donnelley Publishing.

National Fire Protection Association (1986). *Fire Protection Handbook*, 16th ed.

Planer, R. G. (1979). *Fire Loss Control*. Marcel Dekker, New York.

Tuve, R. L. (1976). *Principles of Fire Protection Chemistry*. National Fire Protection Association.

Index

Printed in the United States
by Baker & Taylor Publisher Services